CONTENTS

Acknowledgments

Special thanks to (i)Wikipedia, the free encyclopaedia for some definitions, (ii) department of education's publications of November 2013 – for new specification documents for Mathematics teaching from September 2015.

Iconic Maths was written with clear vision and much conviction that it will enrich the secondary curriculum and empower top achieving students to a higher level of success.

This is an 'extra content' book and should be used in conjunction with any higher mathematics textbook already in existence. The higher mathematics textbooks already in use in schools and colleges will continue to be very useful as **IM** is only an addition to a complete and enriching curriculum. It will complement the existing textbooks. However, the contents of the existing higher mathematics textbooks in our schools and colleges will **not be sufficient** for the new grading system come 2015 for higher attaining students.

Iconic Maths was written by Arinze Edward Oranye, a renowned mathematician and author based in the United Kingdom. A graduate of Petroleum Engineering with master's degree in Mathematics, Arinze attained the position of Head of Mathematics in a State maintained school in the United Kingdom. With his profound experience and dynamism, he has also taken up mentoring for newly qualified teachers and PGCE students at Stewards Academy, Harlow Essex.

First produced in 2014

To order this book, contact Edward on 07825313771, 01279412418 or by email: edward@iconicconcepts.co.uk and edwardoranye@hotmail.com.

Iconic Maths

New GCSE

Higher Tier

Grades 5 - 9

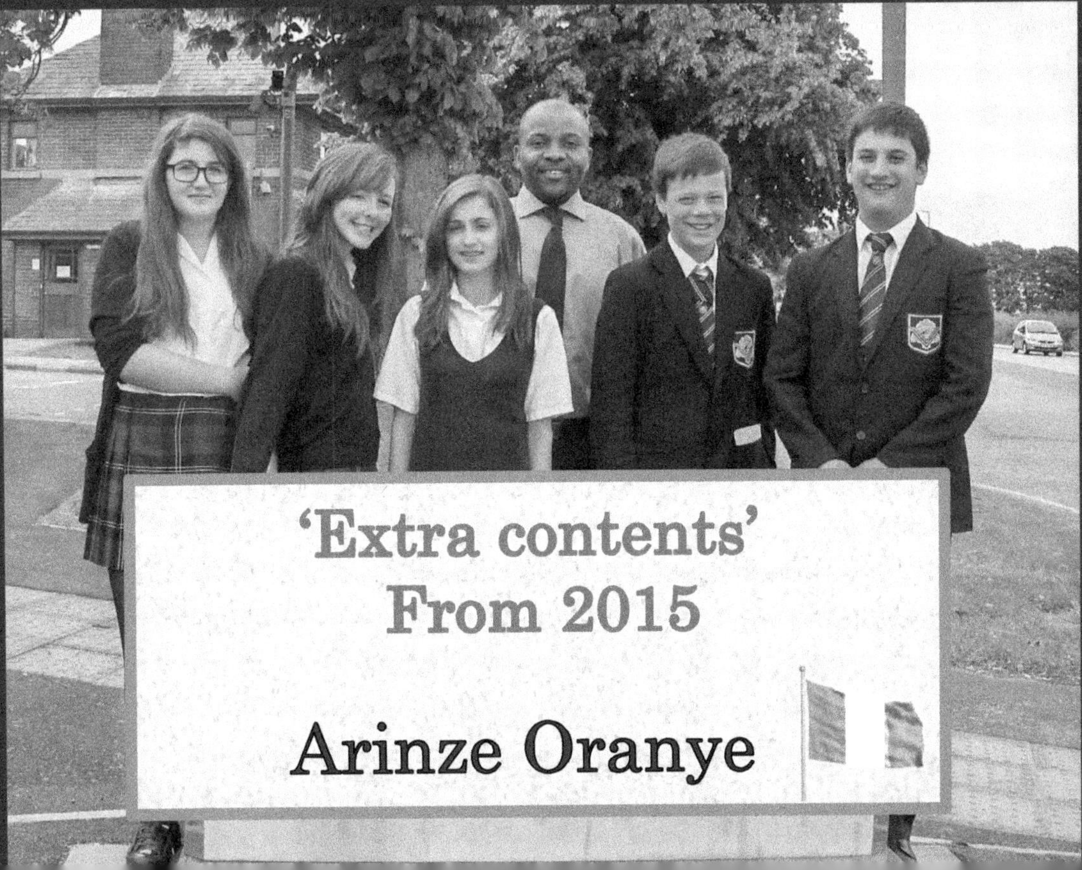

'Extra contents'
From 2015

Arinze Oranye

A **Must-Read** for teachers and organisations

Iconic Maths **(IM)** was developed to accommodate the new GCSE mathematics specifications announced by the government on 1st of November, 2013. The new specifications in GCSE mathematics will be introduced in England for first teaching from September 2015.

This book was written with clear vision and expectations towards high attainment for every top set student with ambition to furthering their education in mathematics. It is an '**extra content'** book and will be used in conjunction with any other higher mathematics book already in existence. There is no need to discard your old textbooks as they will complement the extra content book.

IM was written with the conviction that it will enrich the secondary curriculum and empower top achieving students to a higher level of success.

According to the department of Education, Ref: DFE-00233-2013, schools need to look at what formulae students need to memorise and derive in other to prepare and enrich themselves for GCSE Mathematics examinations in England.

The details below were obtained from the Department of Education publications of 1st November 2013. It is pertaining to new changes and specifications regarding formulae applications in future GCSE Mathematics examinations which will be first taught in England from 2015.

Iconic Maths was written by Edward Arinze Oranye, a renowned mathematician and author based in the United Kingdom. A graduate of Petroleum Engineering with master's degree in Mathematics, Arinze attained the position of Head of Mathematics in a state-maintained school in the United Kingdom. He is currently an experienced Mathematics teacher/mentor at Stewards Academy, Harlow Essex.

Formulae and the New GCSE Mathematics Specifications

1) Students (candidates) are expected to know these formulae. They **must not** be given in the assessment.

- The quadratic formula. The solutions of $ax^2 + bx + c = 0$ where $a \neq 0$

  $$x = \frac{-b \pm \sqrt{b^2 - 4ac}}{2a}$$

  where r = radius and d = diameter of the circle

- Area of a circle = πr^2
- Circumference = πd or $2\pi r$
- Pythagoras's theorem

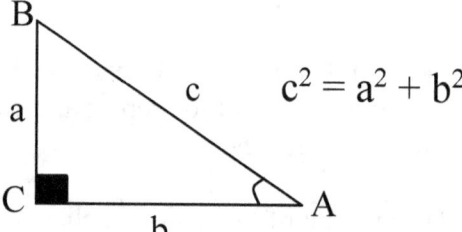

$$c^2 = a^2 + b^2$$

Fig 1

- Trigonometry formulae
  From figure 1 above, $\text{Sin } A = \dfrac{a}{c}$ $\text{Cos } A = \dfrac{b}{c}$ $\text{Tan } A = \dfrac{a}{b}$

- Sine rule

  $$\frac{a}{\text{Sin } A} = \frac{b}{\text{Sin } B} = \frac{c}{\text{Sin } C}$$

- Cosine rule $a^2 = b^2 + c^2 - 2bc \text{ CosA}, \quad b^2 = a^2 + c^2 - 2ac \text{ CosB}$
  $$c^2 = a^2 + b^2 - 2ab \text{ CosC}$$
- Area of triangle = ½ bc SinA
  = ½ ac SinB
  = ½ ab SinC

  where a, b and c are the lengths of the sides.

2) Formulae not specified in the content but **should be derived** or informally understood by students. These formulae **must not** be given in the examination.

- Perimeter, area, surface area and volume formulae

- Area of trapezium

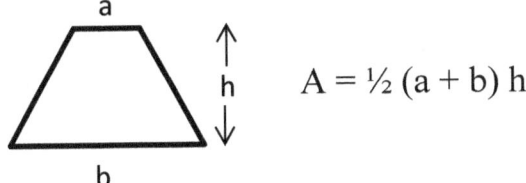

$$A = ½ (a + b) h$$

- Volume of any prism = area of cross section × length

- Compound Interest

$$\text{Total accrued} = P \left(1 + \frac{r}{100}\right)^n$$

where P = principal amount
r = interest rate over a given period
n = number of times that the interest is compounded.

- Probability
where P(A) = Probability of outcome A, and P(B) is the probability of outcome B

$$P(A \text{ or } B) = P(A) + P(B) - P(A \text{ and } B)$$

$$P(A \text{ and } B) = P(A \text{ given } B). P(B)$$

3) Formulae to be used but **need not** memorise. They can be given in examinations either in relevant question or in a list which candidates select and apply as appropriate.

- 3 – D shapes
  Curved surface area of a cone = $\pi r \ell$
  Surface area of sphere = $4\pi r^2$
  Volume of a sphere = $4/3\ \pi r^3$
  Volume of a cone = $1/3\ \pi r^2 h$

  where
  r = the radius of the sphere or cone
  ℓ = slant height of a cone
  h = perpendicular height of a cone

- Kinematics formulae
  $v = u + at$
  $s = ut + \frac{1}{2}at^2$
  $v^2 = u^2 + 2as$

  where
  a = constant acceleration
  u = initial velocity
  v = final velocity
  s = displacement from the position when t = 0
  t = time taken

Source: Department of Education publications of 1st of November 2013.

1 More than two binomials

This section covers the following topics:

- Multiplying three brackets
- Expanding a squared bracket and another bracket

LEARNING OBJECTIVES

By the end of this unit, you should be able to:

- Use the knowledge of expanding double brackets to multiply three brackets
- Be confident with algebraic manipulations

KEYWORDS

- Binomial
- Expansion
- Brackets

1.1 MORE THAN TWO BINOMIALS

Expressions that contain two terms are called **binomials**.

The expression **a + 5** contains two terms, a and +5, hence it is called a binomial.

In this section, we shall consider multiplying more than two binomials.

Example 1: Expand and simplify $(x + 2)(x + 3)(x + 4)$

There are lots of ways we can multiply these binomials. However, only two methods shall be considered - the "FOIL" method and the "GRID" method.

FOIL method

F – first
O – outside
I – inside
L – last

This means to multiply the first terms in each bracket followed by the outside terms, then the inside terms and finally, the last terms.

Multiply the first two brackets $(x + 2)(x + 3)$ using the foil method.

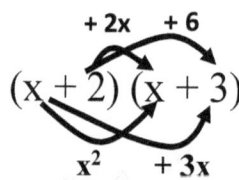

$$+ 2x \quad + 6$$
$$(x + 2)(x + 3)$$
$$x^2 \quad + 3x$$

$x \times x = \mathbf{x^2}$, $x \times 3 = \mathbf{3x}$, $2 \times x = \mathbf{2x}$ and $2 \times 3 = \mathbf{6}$

Add all the terms will give $x^2 + 3x + 2x + 6$

Collecting like terms: $x^2 + 5x + 6$

Now, multiply by the remaining bracket $(x + 4)$.

$(x^2 + 5x + 6)(x + 4)$

$x^2 \times x = \mathbf{x^3}$, $x^2 \times 4 = \mathbf{4x^2}$

$5x \times x = \mathbf{5x^2}$, $5x \times 4 = \mathbf{20x}$

$6 \times x = \mathbf{6x}$, and $6 \times 4 = \mathbf{24}$

Collecting like terms:

$x^3 + 4x^2 + 5x^2 + 20x + 6x + 24$

$= \mathbf{x^3 + 9x^2 + 26x + 24}$ ✔

GRID method

First multiply $(x + 2)$ and $(x + 3)$

×	x	2
x	x^2	2x
3	3x	6

$x^2 + 3x + 2x + 6 = x^2 + 5x + 6$

Then, multiply by $(x + 4)$

x	x^2	5x	6
x	x^3	$5x^2$	6x
4	$4x^2$	20x	24

$x^3 + 4x^2 + 5x^2 + 20x + 6x + 24$

$= x^3 + 9x^2 + 26x + 24$ ✔

Example 2

Expand and simplify $(3x - 2)(x - 4)^2$

Solution: Expand $(x - 4)^2$ and then multiply by $(3x - 2)$

$(x - 4)^2 = (x - 4)(x - 4)$

Using the grid method:

x	x	-4
x	x^2	-4x
-4	-4x	16

$= x^2 - 4x - 4x + 16$
$= x^2 - 8x + 16$

Next, multiply by $(3x - 2)$

x	x^2	-8x	16
3x	$3x^3$	$-24x^2$	48x
-2	$-2x^2$	+ 16x	-32

$= 3x^3 - 2x^2 - 24x^2 + 16x + 48x - 32$
$= \mathbf{3x^3 - 26x^2 + 64x - 32}$ ✓

EXERCISE 1A

1) Expand and simplify.

a) $(x + 1)(x + 2)(x + 4)$
b) $(x + 2)(x + 3)(x + 1)$
c) $(x - 2)^2 (x - 5)$
d) $(b - c)(c + 3)(d - 2)$
e) $(w^2 - 3w + 5)(w - 7)$

2) Expand and simplify.

a) $(2x + 6)(3x + 2)(5x + 1)$
b) $(5x - 4)^3$

3) a) Work out the expression for the volume of the cuboid below.

b) Work out the expression for the surface area of the cuboid.

All lengths in cm.

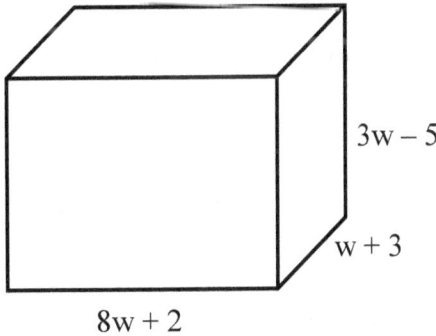

3w − 5

w + 3

8w + 2

c) If the volume is 3360 cm³, work out the dimensions of the cuboid.

4) $(2c + 1)(c - 1)(3c + 2) = gc^3 + c^2 - 5c - 2.$

Find the value of g and c.

2 Quadratic Equations

This section covers the following topics:

- Factorising quadratic expressions

- Solving quadratic equations

LEARNING OBJECTIVES

By the end of this section, you should be able to:

- Solve quadratic equations by factorisation

- Solve quadratic equations by using the quadratic formula

- Solve quadratic equations by completing the square method

- Identify turning points and equations of the symmetry

- Sketch quadratic graphs

- Understand the Discriminant

KEYWORDS

- Quadratic equation
- Completing the square
- Turning points
- Discriminant
- Quadratic formula

Note: Students are expected to memorise the quadratic formula.

2.1 SOLVING QUADRATIC EQUATIONS

By now, you must be familiar with factoring quadratics. Some expressions will factorise while some will not. In this section, we shall consider solving quadratic equations by factorisation method, quadratic formula and by completing the square.

BY FACTORISATION

Example 1: Solve $x^2 - 4 = 0$
Factorise the left side.
$(x - 2)(x + 2) = 0$
Either $(x - 2) = 0$ or $(x + 2) = 0$
When $x - 2 = 0$, $x = 2$
When $x + 2 = 0$, $x = -2$.

Therefore, the solutions are
x = 2 and x = -2

Example 2: Solve $5x^2 - 45 = 0$

Add 45 to both sides
$5x^2 = 45$
Divide both sides by 5
$x^2 = 9$
$x = \sqrt{9}$
x = ± 3

Example 3:
Solve $x^2 + 12x + 35 = 0$

Factorise the left side
$(x + 5)(x + 7) = 0$
From $x + 5 = 0$, $x = -5$
From $x + 7 = 0$, $x = -7$
The solutions are **x = -5 and x = -7**.

Example 4: Solve $5x^2 + 16x + 3 = 0$

Factorise the left-hand side
$(x + 3)(5x + 1) = 0$

From $x + 3 = 0$, $x = -3$
From $5x + 1 = 0$, $x = -\dfrac{1}{5}$

The solutions are **x = -3** and $\mathbf{x = -\dfrac{1}{5}}$

BY QUADRATIC FORMULA

The quadratic formula is used where a quadratic expression cannot be factorised.

The quadratic equation is in the form $ax^2 + bx + c = 0$ where $a \neq 0$. The quadratic formula

$$x = \frac{-b \pm \sqrt{b^2 - 4ac}}{2a}$$

gives the solutions (roots) of the quadratic equation.

Example 1:
Solve the equation $x^2 + 5x + 5 = 0$

Solution: The expression cannot be factorised, so we use the quadratic formula.

Comparing $\mathbf{1x^2 + 5x + 5 = 0}$ with

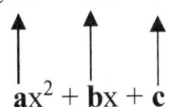

$$ax^2 + bx + c$$

gives a = 1, b = 5 and c = 5

Next is to substitute the values of a, b and c into the quadratic formula

$$x = \frac{-b \pm \sqrt{b^2 - 4ac}}{2a}$$

$$x = \frac{-(5) \pm \sqrt{(5)^2 - 4 \times 1 \times 5}}{2 \times 1}$$

$$x = \frac{-5 \pm \sqrt{25 - 20}}{2}$$

$$x = \frac{-5 \pm \sqrt{5}}{2}$$

$$x = \frac{-5 + \sqrt{5}}{2} \quad \text{or} \quad x = \frac{-5 - \sqrt{5}}{2}$$

$$x = -1.38 \qquad \text{or} \quad x = -3.62$$

General curve of a quadratic equation.

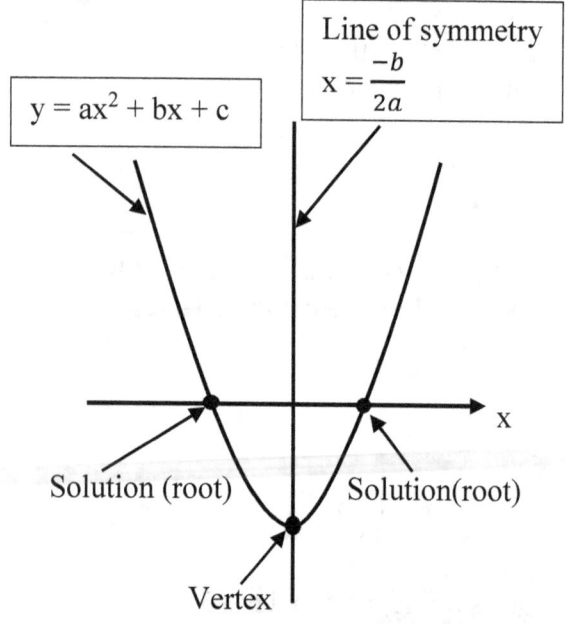

$y = ax^2 + bx + c$

Line of symmetry
$$x = \frac{-b}{2a}$$

Solution (root)

Solution(root)

Vertex

Example 2: Solve the quadratic equation $2x^2 - 2x - 5 = 0$ and round to 2 decimal places.

Solution: a = 2, b = -2 and c = -5
Using the quadratic equation

$$x = \frac{-b \pm \sqrt{b^2 - 4ac}}{2a}$$

$$x = \frac{-(-2) \pm \sqrt{(-2)^2 - 4 \times 2 \times -5}}{2 \times 2}$$

$$x = \frac{2 \pm \sqrt{4 + 40}}{4}$$

$$x = \frac{2 \pm \sqrt{44}}{4}$$

$$x = \frac{2 + \sqrt{44}}{4} \quad \text{or} \quad \frac{2 - \sqrt{44}}{4}$$

$$x = 2.16 \qquad \text{or } x = -1.16$$

Example 3:
Sole the equation $x(5 + x) = 7$ and round to 2 decimal places.

Solution: Expand the brackets.
$5x + x^2 = 7$
Subtract 7 on both sides
$5x + x^2 - 7 = 0$
Rearrange in the form $ax^2 + bx + c$
$x^2 + 5x - 7 = 0$
a = 1, b = 5 and c = -7
Using the quadratic formula

$$x = \frac{-5 \pm \sqrt{25 - (4 \times 1 \times -7)}}{2 \times 1}$$

$$x = \frac{-5 \pm \sqrt{25 + 28}}{2} = \frac{-5 \pm \sqrt{53}}{2}$$

$$x = \frac{-5 + \sqrt{53}}{2} \quad \text{or } x = \frac{-5 - \sqrt{53}}{2}$$

x = 1.14 or x = -6.14

Example 4: Solve $3x(x - 2) = (x + 2)^2 - 5$

Solution: Expand the brackets
$3x^2 - 6x = x^2 + 4x + 4 - 5$
$3x^2 - 6x - x^2 - 4x - 4 + 5 = 0$
$2x^2 - 10x + 1 = 0$

Using the quadratic formula when
a = 2, b = -10 and c = 1,

$$x = \frac{-b \pm \sqrt{b^2 - 4ac}}{2a}$$

$$x = \frac{-(-10) \pm \sqrt{(-10)^2 - 4 \times 2 \times 1}}{2 \times 2}$$

$$x = \frac{10 \pm \sqrt{100 - 8}}{4}$$

$$x = \frac{10 \pm \sqrt{92}}{4}$$

$$x = \frac{10 + \sqrt{92}}{4} \quad \text{or} \quad x = \frac{10 - \sqrt{92}}{4}$$

x = 4.90   or **x = 0.10**

Please note: In surd form, the roots are

$$x = \frac{10 + \sqrt{92}}{4} \quad \text{or} \quad x = \frac{10 - \sqrt{92}}{4}$$

EXERCISE 2A

1) Solve the quadratic equations by factorisation.

a) $5x^2 + 16x + 3 = 0$
b) $2x^2 + 15x + 28 = 0$
c) $10x^2 + 23x + 12 = 0$
d) $4x^2 + 41x + 10 = 0$
e) $6x^2 + 31x + 5 = 0$

2) Solve the quadratic equations.

a) $8x^2 - 10x - 12 = 0$
b) $6x^2 - 19x + 15 = 0$
c) $4x^2 - 13x + 10 = 0$
d) $42x^2 + 17x - 15 = 0$
e) $30x^2 + 20x - 10 = 0$

3) Solve the quadratic equations.

a) $a^2 - 4 = 0$ f) $r^2 - 1 = 0$
b) $a^2 - 9 = 0$ g) $k^2 - 81 = 0$
c) $x^2 - 100 = 0$ h) $w^2 - 169 - 0$
d) $x^2 - 144 = 0$ i) $w^2 - 324 = 0$
e) $x^2 - 225 = 0$ j) $m^2 - 729 = 0$

4) Solve the equations below.

a) $w^2 - 6w - 55 = 0$
b) $w^2 - 8w - 48 = 0$
c) $w^2 + 4x - 60 = 0$
d) $w^2 - 16x + 64 = 0$
e) $w^2 - 13x + 42 = 0$

EXERCISE 2B

Solve the quadratic equations using the quadratic formula. Leave your answers in surd form where possible.

1) $x^2 + 9x + 2 = 0$
2) $x^2 + 4x + 1 = 0$
3) $x^2 + 7x - 6 = 0$
4) $x^2 + 5x - 3 = 0$
5) $x^2 - 3x - 2 = 0$
6) $2x^2 + 7x + 3 = 0$
7) $3x^2 + 8x - 2 = 0$
8) $6x^2 - 5x - 3 = 0$
9) $4x^2 + 5x - 3 = 0$
10) $x^2 - x - 7 = 0$

EXERCISE 2C

1) Round your answers from exercise 17F question 3 to 3 significant figures.

2) Use the quadratic formula to solve the equations below.
a) $x^2 = x + 7$ d) $y(y - 5) = 3$
b) $2x^2 = 11 - x$ e) $5x(x - 3) = (x + 3)^2 - 7$
c) $x(x + 7) = -9$ f) $(y - 9)^2 = 11$

3) Th difference between two numbers is 7. When multiplied together, the answer is 5. Work out the two numbers and leave your answers to 2 decimal places.

4) Find two consecutive whole numbers whose product is 306.

5) The base of a right-angled triangle is 10m longer than the perpendicular height. The area of the triangle is 37.5 m^2. Work out the perpendicular height.

2.2 THE DISCRIMINANT

In the quadratic formula

$$x = \frac{-b \pm \sqrt{b^2 - 4ac}}{2a},$$

the expression under the square root sign $b^2 - 4ac$ is called the **discriminant**.

Knowledge of the Discriminant

1) $\boxed{b^2 - 4ac > 0}$

The quadratic equation has two real roots (solutions) if $b^2 - 4ac$ is greater than zero (0). The parabola which is the shape of the quadratic curve cuts the x-axis at two distinct places. Consider the parabola represented by the equation $x^2 + 5x - 7 = 0$.

The discriminant $b^2 - 4ac$
$= 5^2 - 4 \times 1 \times -7 = 25 + 28 = \textbf{53}$

This is greater than zero. It means that there are two real roots (solutions) and the curve cuts the x-axis at two points.

To show this, solve the quadratic equation. $a = 1$, $b = 5$ and $c = -7$.

$$x = \frac{-5 \pm \sqrt{5^2 - (4 \times 1 \times -7)}}{2 \times 1}$$

$x = \textbf{1.14}$ or $x = \textbf{- 6.14}$

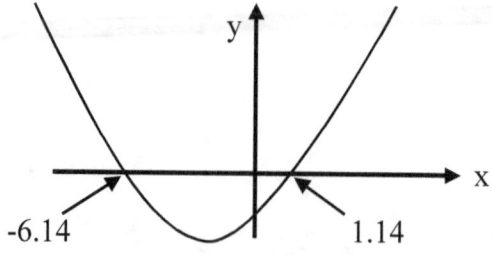

2) $\boxed{b^2 - 4ac < 0}$

The quadratic equation has **no real roots** or solutions if the discriminant $b^2 - 4ac$ is less than zero. This is because the square root of a negative number does not exist (will give a math error on a calculator) and therefore, the parabola **does not** cut the x-axis.

Consider the quadratic equation $2x^2 - 3x + 5 = 0$, where a = 2, b = -3 and c = 5. The discriminant $b^2 - 4ac$ $= (-3^2) - 4 \times 2 \times 5 = 9 - 40 = \textbf{-31}$

When used in the quadratic equation, it becomes $\sqrt{-31}$. The square root of a negative number is undefined and does not exist. Therefore, the curve of the quadratic equation **will not** cut the x-axis and will have no solution.

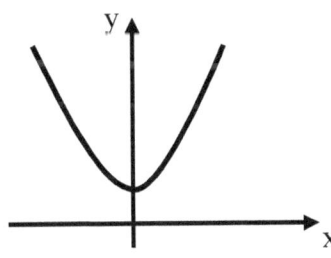

3) $\boxed{b^2 - 4ac = 0}$

When the discriminant $b^2 - 4ac = 0$, the x-axis is the horizontal tangent to the parabola (curve). The quadratic equation has only **one root** (solution). Consider the quadratic equation $3x^2 + 6x + 3 = 0$ where a = 3, b = 6 and c = 3.

Using the discriminant $b^2 - 4ac$, $6^2 - (4 \times 3 \times 3) = 36 - 36 = \textbf{0}$ Applying this in the quadratic formula gives:

$$x = \frac{-b \pm \sqrt{0}}{2a} \quad = \quad \frac{-b}{2a}$$

It is now evident that when the discriminant = 0, the part of the quadratic equation that contains the square root sign becomes non-existent as the square root of zero is 0. Therefore, we have only one root or solution. If we continue solving the quadratic equation above, $x = \frac{-6}{6} = -1$, showing only one root at (-1, 0).

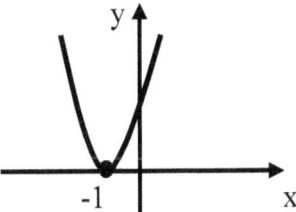

Example 1: Using only the discriminant, decide whether the quadratic equation $2x^2 + 3x - 1 = 0$ has two, one or no solutions.

Solution: a = 2, b = 3 and c = -1. Using the discriminant $b^2 - 4ac$, $= 3^2 - (4 \times 2 \times -1) = 9 + 8 = 17$ Since **17 > 0**, there are two solutions.

Example 2:
Decide whether $5x^2 - 2x + 3 = 0$ has one, two or no solutions.
Solution: $b^2 - 4ac = (-2)^2 - (4 \times 5 \times 3)$ $= 4 - 60 = -56$.
Since **-56 < 0**, there are no solutions.

EXERCISE 2D

For questions **1 – 8**, show whether the quadratic equations below have one, two or no solutions. Sketch the graphs.

1) $x^2 + 3x + 6 = 0$ 2) $3x^2 + 5x - 9 = 0$

3) $2x^2 - 2x - 7 = 0$ 4) $3x^2 + 6x + 3 = 0$

5) $5x^2 + 10x + 5 = 0$ 6) $x^2 - 4x + 5 = 0$

7) $x^2 + 3x + 2 = 0$ 8) $6x^2 + 9x - 3 = 0$

9) **a** Work out the coordinates of P and Q
   b Write down the equation of the line of symmetry.

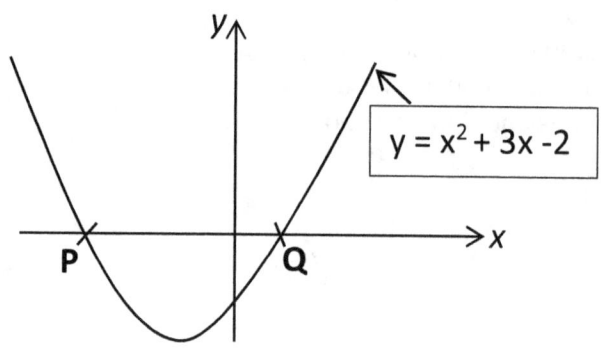

$y = x^2 + 3x - 2$

2.3 COMPLETING THE SQUARE

Completing the square is an important technique for solving quadratic equations.
The maximum and minimum **turning points** of the parabola can be found using the method of completing the squares.
Consider the function $y = x^2 + 7x - 3$

From graphs of a quadratic function, where the coefficient of x^2 is positive, any of the graphs below is possible.

A
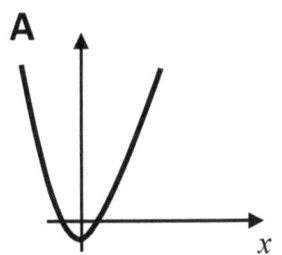
B
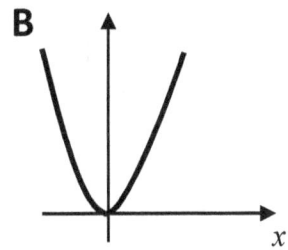
C
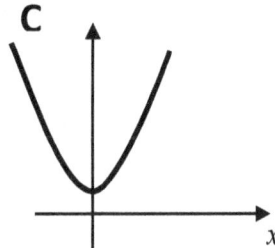

We may be interested in finding the coordinates of the minimum point. The parabola will have a minimum (turning) point on the graph and we may be interested in finding the x and y values at this point. Is the turning point below or above the x- axis?
We may use completing the square technique to answer the question.

What is the meaning of complete or perfect square?

An expression of the form $(x + a)^2$ is called a **complete** or **perfect** square.
Expanding will give $(x + a)^2 = (x + a)(x + a) = x^2 + 2ax + a^2$

Also, $(x - a)^2$ is a complete square. Expanding the bracket will give
$(x - a)^2 = (x - a)(x - a) = x^2 - 2ax + a^2$

Consider the expression $(x + 3)^2$

$(x + 3)^2 = (x + 3)(x + 3)$

$\quad\quad = x^2 + 6x + 9$

$x^2 + 6x + 9$ is a perfect square

Methods of completing the square

When the coefficient of x^2 is 1

Before we solve quadratic equations using completing the square method, we must first write the expression in completed square form. It must be written in a form which has a square term and a constant.

Method 1: General method

Example 1: Write $x^2 + 6x + 3$ in the form $(x + p)^2 + q$
Consider the expressions $\mathbf{x^2 + 6x}$.

Halve the coefficient of x in the expression 6x. This is 3.
As a perfect square, $(x + 3)^2 = x^2 + 6x + 9$

To get back $x^2 + 6x$, we subtract 9
$x^2 + 6x + 9 - \mathbf{9} + 3$
but $(x + 3)^2 = x^2 + 6x + 9$

Therefore, $(x + 3)^2 - 9 + 3$
$= (x + 3)^2 - 6$

Finally, $x^2 + 6x + 3$ in the form $(x + p)^2 + q = \mathbf{(x + 3)^2 - 6}$

We shall look at turning points at a later stage and its implications on the quadratic graph.

Also, we can solve quadratic equations using the completed square form. All these will be dealt with at a later stage in this chapter.

Method 2: The co-efficient method

This method of completing the square compares the coefficients of the original quadratic to those of the multiplied out completed form.

Example 1: Write $x^2 + 6x + 3$ in the form $(x + p)^2 + q$

Expanding gives $\longrightarrow$ $(x^2 + 2px + p^2) + q$

This compares to:

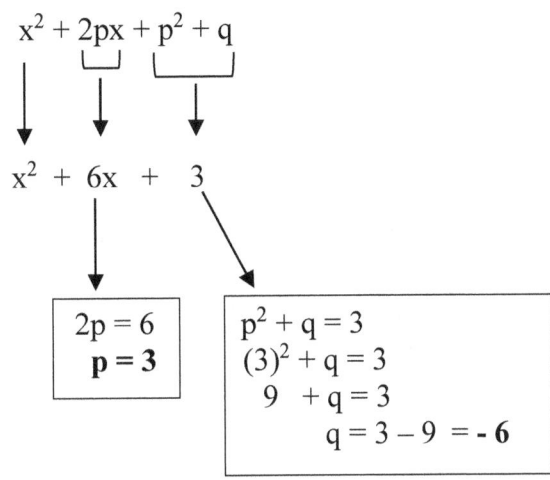

$$2p = 6$$
$$p = 3$$

$$p^2 + q = 3$$
$$(3)^2 + q = 3$$
$$9 + q = 3$$
$$q = 3 - 9 = -6$$

Putting the values of p and q into the form $(x + p)^2 + q$ will give: $(x + 3)^2 - 6$ ✓

EXERCISE 2E

Write in completed square form the expressions below using any method.

1) $x^2 + 8x + 2$

2) $x^2 + 10x + 7$

3) $x^2 + 6x - 9$

4) $x^2 - 6x - 1$

5) $x^2 - 16x + 5$

6) $x^2 + 2x - 5$

7) $x^2 + 5x + 5$

8) $x^2 - 7x - 9$

Write the following expressions in the form $(a + b)^2 + c$. Write down the values of b and c.

9) $x^2 + 20x + 7$

10) $x^2 - 7x + 5$

11) $x^2 - 3x - 6$

12) $x^2 + 14x + 8$

13) $x^2 + 9x + 2$

14) $x^2 - 4x - 4$

15) $x^2 + 2x + 5$

16) $x^2 + 6x + 13$

17) Work out the values of **p** and **q** such that $x^2 + 8x - 8 = (x + p)^2 - q$

2.4 SOLVING QUADRATIC EQUATIONS BY THE METHOD OF COMPLETING THE SQUARE

Example 1: Solve the equation $x^2 - 6x + 3 = 0$ by completing the square.

$(x - 3)^2 = x^2 - 6x + 9$
$(x^2 - 6x + 9) - 9 + 3 = 0$
$(x - 3)^2 - 6 = 0$
Add 6 to both sides
$(x - 3)^2 = 6$
$x - 3 = \sqrt{6}$
$x - 3 = \pm \sqrt{6}$
Add 3 to both sides
$x = 3 \pm \sqrt{6}$ .... this is the answer in surd form. It is the exact answer.

However, we may proceed to evaluate the whole answer.
$x = 3 + 2.449....$ $x = 5.449...$
or
$x = 3 - 2.449...$ $x = 0.5505...$

By using completing the square method, the solutions to the quadratic equation
$x^2 - 6x + 3 = 0$ is **$x = 5.449...$ or $x = 0.5505...$**

Example 2:
Solve the equation $n^2 + 5n - 2 = 0$ using the completing the square method.
Leave your answer in surd form.

Solution:
$5 \div 2 = 2.5$
$(n + 2.5)^2 = n^2 + 5n + 6.25$
$(n^2 + 5n + 6.25) - 6.25 - 2 = 0$
$(n + 2.5)^2 - 8.25 = 0$

Add 8.25 to both sides
$(n + 2.5)^2 = 8.25$
$n + 2.5 = \pm \sqrt{8.25}$

Subtract 2.5 from both sides
$n = \pm \sqrt{8.25} - 2.5$

$n = \sqrt{8.25} - 2.5$ or $n = -\sqrt{8.25} - 2.5$

EXERCISE 2F

1) Solve the equations by completing the square. Leave your answers in surd form where appropriate.

a) $m^2 + 8m - 3 = 0$ b) $x^2 + 8x - 5 = 0$
c) $x^2 - 4x - 1 = 0$ d) $x^2 + 10x + 6 = 0$
e) $x^2 - 6x - 2 = 0$ f) $x^2 + 12x + 9 = 0$

2) Solve the following equations by completing the square. Leave your answers to 2 decimal places where possible.

a) $c^2 + 2c - 3 = 0$ b) $x^2 + 8x - 5 = 0$
c) $x^2 - 4x - 1 = 0$ d) $x^2 + 10x + 6 = 0$
e) $x^2 - 6x - 2 = 0$ f) $y^2 + 12y + 9 = 0$

2.5 When the coefficient of x^2 is not 1

When the coefficient of x^2 is not 1, take out the coefficient as a **factor** and complete the squares as usual.

Example 1: Write $5x^2 + 10x + 2$ in the form **a(x + p)² + q.**

The coefficient of x^2 is not 1. It is 5.

Take out **5** as a factor.

$$5x^2 + 10x + 2 \quad = 5\left[x^2 + 2x + 2/5\right]$$

$$= 5\left[(x + 1)^2 - 3/5\right]$$

Multiply out the brackets

$$= 5(x + 1)^2 - 3$$

Therefore, a = 5, p = 1 and q = -3

Example 2: Write $\frac{1}{4}x^2 - 5x + 4$ in the form $a(x + p)^2 + q$

Take out $\frac{1}{4}$ as a factor.

$$\frac{1}{4}x^2 - 5x + 4 = \frac{1}{4}\left[x^2 - 20x + 16 \right]$$

$$= \frac{1}{4}\left[(x - 10)^2 - 84 \right]$$

Multiply out the brackets
$$= \frac{1}{4}(x - 10)^2 - 21$$

EXERCISE 2G

Write the expressions in the form $a(x + p)^2 + q$

1) $3x^2 + 6x + 9$ 2) $5x^2 + 15x + 10$
3) $4x^2 - 6x + 1$ 4) $7x^2 - 14x + 6$
5) $\frac{1}{2}n^2 + 5n + 6$ 6) $\frac{3}{4}k^2 - 3k + 6$

Example 3: Solve the quadratic equation $5x^2 + 10x + 2 = 0$ by completing the square.

Look back to example 1. In completed square form, $5x^2 + 10x + 2 = \mathbf{5(x + 1)^2 - 3}$
Therefore, $5(x + 1)^2 - 3 = 0$
             add 3 to both sides
$5(x + 1)^2 = 3$
             divide both sides by 5
$(x + 1)^2 = 3/5 = 0.6$
$x + 1 = \pm \sqrt{0.6}$
             subtract 1 from both sides

$x = \sqrt{0.6} - 1$ or $x = -\sqrt{0.6} - 1$

$x = -0.23$ to 2 dp   or $x = -1.77$ to 2dp.

EXERCISE 2H

Solve the equations using completing the square method to 2 decimal places where appropriate.
1) $5x^2 + 15x + 10 = 0$ 2) $4x^2 - 6x + 1 = 0$
3) $7x^2 - 14x + 6 = 0$ 4) $\frac{1}{2}n^2 + 5n + 6 = 0$

2.6 TURNING POINT

The vertex of a parabola (quadratic curve) is the place where it **turns**. It is therefore called the turning point. The perpendicular line that passes through the vertex is also known as the axis of symmetry of the parabola.

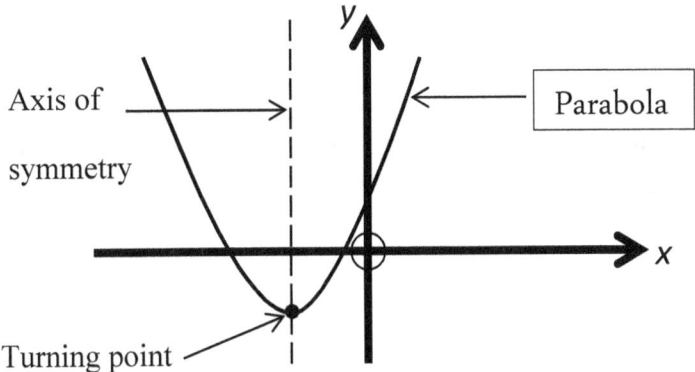

When equations are written in the form **y = a(x + p)² + q**, the quadratic graph formed will have axis of symmetry x = - p and turning point (-p, q).

Example 1: **a)** Write $3x^2 + 6x + 7$ in the form **a(x + p)² + q**
  b) Find the turning point of the graph
  c) Write the equation of the line of symmetry

Solution:
a) Factor out 3 in $3x^2 + 6x + 7$ $\quad = 3\left[(x^2 + 2x + 7/3)\right]$

$$= 3\left[(x + 1)^2 + 4/3\ \right]$$

$$= \mathbf{3\ (x + 1)^2 + 4}$$

b) The minimum value is obtained when the expression in the bracket is equal to zero. This happens when x = -1. Therefore, the turning point occurs at (-1, 4)

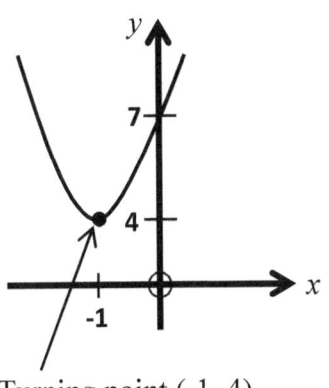

Turning point (-1, 4)

c)

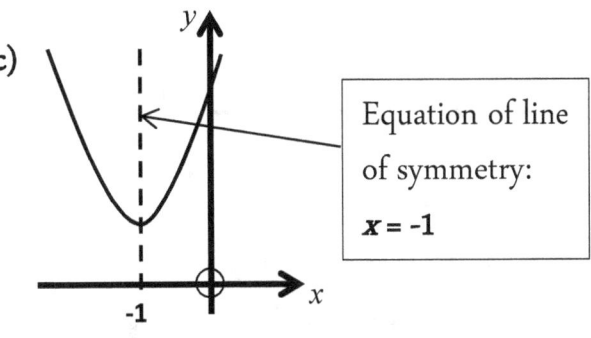

Equation of line of symmetry:

x = -1

25

2.7 SKETCHING THE GRAPH OF A QUADRATIC EQUATION IN COMPLETED SQUARE FORM

$5x^2 + 10x + 2 = 5(x + 1)^2 - 3$ in completed square form.

1) The shape of the graph is U- shaped, since it has a positive gradient (5).
2) The turning point is **(-1, -3)** and the axis of symmetry is the line x = -1.
3) The minimum value is y = -3
4) The y – intercept occurs when x = 0. In this case, it is **y = 2**
5) The solutions of the quadratic equation are where the graph cuts the x – axis. This occurs when y = 0. We need to solve the equation to find out the two points.

Therefore, $5(x + 1)^2 - 3 = 0$
              add 3 to both sides
$5(x + 1)^2 = 3$
              divide both sides by 5
$(x + 1)^2 = 3/5 = 0.6$
$x + 1 = \pm \sqrt{0.6}$
              subtract 1 from both sides
$x = \sqrt{0.6} - 1$ or $x = - \sqrt{0.6} - 1$
x = - 0.23 to 2 dp   or **x = -1.77** to 2dp.

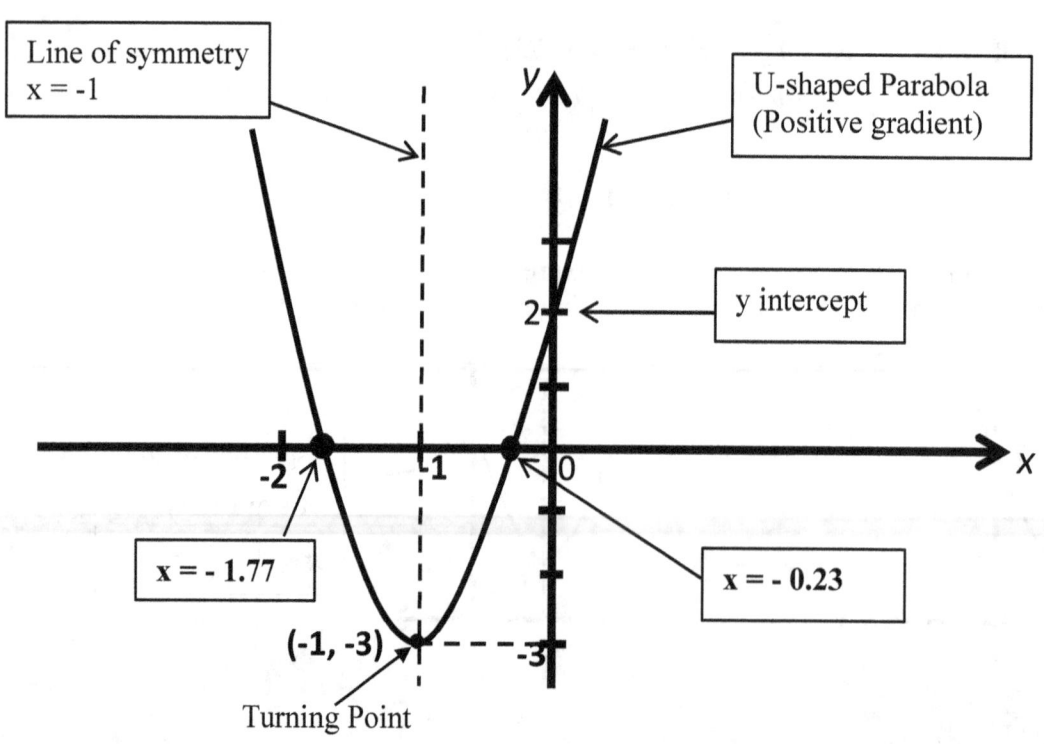

Line of symmetry
x = -1

U-shaped Parabola
(Positive gradient)

y intercept

x = - 1.77

x = - 0.23

(-1, -3)

Turning Point

26

EXERCISE 2I

Write the quadratic equations in the form $a(x + p)^2 + q$

1) $2x^2 + 6x + 7$

2) $3x^2 - 9x + 1$

3) $4b^2 + 4b + 3$

4) $7c^2 - 21c + 8$

5) $5x^2 - 10x - 5$

6) $9x^2 + 9x - 5$

7) In questions 1 – 6 above, work out the values of a, p and q.

8) Sketch the graphs of (i) $2x^2 + 6x + 7$

(ii) $3x^2 - 9x + 1$

9) Write down (i) the turning points in questions 1 - 6

(ii) the equation of the line of symmetry in questions 1 - 6

10) a Form a quadratic equation using the lengths of the triangle below

b Find the solutions of this equation by completing the squares.

c Write down the lengths of the triangles in centimetres to 1 decimal place.

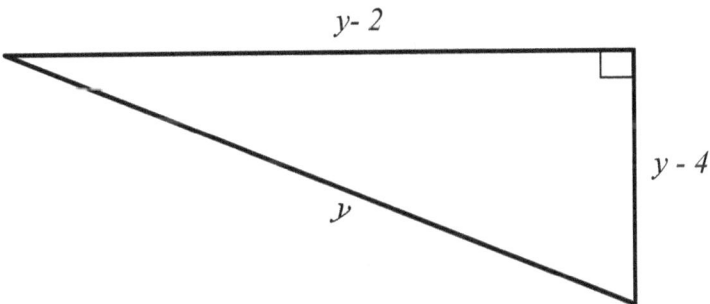

11) Solve the quadratic equations below.

a) $2x^2 + 6x + 7 = 0$

b) $3x^2 - 9x + 1 = 0$

c) $4b^2 + 4b + 3 = 0$

d) $7c^2 - 21c + 8 = 0$

e) $5x^2 - 10x - 5 = 0$

f) $9x^2 + 9x - 5 = 0$

12) a Find c and d such that $3 + 6x - 3x^2 = d - (x - c)^2$

b Solve the quadratic equation $3 + 6x - 3x^2 = 0$

c Write down the turning point on the graph.

d Equation of line of symmetry

3 Iterations

This section covers the following topics:

- Process of iteration
- Solutions using iterations

LEARNING OBJECTIVES

By the end of this unit, you should be able to:

a) Find approximate solutions to equations numerically using iterations

b) Understand the process of iteration.

KEYWORDS

- Iteration

- Starting point

- Iterative formula

3.1 Iterative process

An **iterative method** is a mathematical procedure that generates a sequence of improving approximate solutions for a class of problems.

Iteration is an act of repeating a process with the aim of approaching a desired goal, target or results. Each repetition of the process is also called *iteration.*

An iterative method is called *convergent* if the corresponding sequence converges for given initial approximations.

In the process of finding the root of an equation (or a solution of a system of equations), an iterative method uses an initial guess to generate successive approximations to a solution.

In contrast, direct methods attempt to solve the problem by finite sequence of operations.

Example 1: Solve the equation $x^3 - 3x + 5 = 0$ using iteration method, correct to 2 dp.

Steps to iteration

Step1: The equation must be *re-arranged* to form an iterative formula.
$$x^3 - 3x + 5 = 0$$

add 3x to both sides

$$x^3 + 5 = 3x$$

subtract 5 from both sides

$$x^3 = 3x - 5$$

$$x = \sqrt[3]{3x - 5}$$

The iterative formula is

$$x_{n+1} = \sqrt[3]{3x_n - 5}$$

Step 2: Guess or choose the starting value. This could be given or not.

Let the **starting** value be 5. Therefore, $x_1 = 5$

Step 3: Substitute the value of x_1 into the iterative formula

$$x_{n+1} = \sqrt[3]{3x_n - 5}$$

Use $x_1 = 5$ in the iterative formula above to find x_2

$$x_2 = \sqrt[3]{(3 \times 5) - 5} \qquad = 2.1544\ldots\ldots$$

$$x_3 = \sqrt[3]{(3 \times 2.154\ldots\ldots) - 5} = 1.135\ldots\ldots$$

Continue with the iteration process until the values tend to converge at a particular number. That is the required degree of accuracy.

$$x_4 = -1.168\ldots\ldots$$
$$x_5 = -2.041\ldots\ldots$$
$$x_6 = -2.232\ldots\ldots$$
$$x_7 = -2.269\ldots\ldots$$
$$x_8 = -2.277\ldots\ldots$$
$$x_9 = -2.278\ldots\ldots$$
$$x_{10} = -2.278\ldots\ldots$$
$$x_{11} = -2.279\ldots\ldots$$
$$x_{12} = -2.279\ldots\ldots$$

There is no point continuing the iteration process at this point as the values converge to -2.28 to 2 decimal places. Therefore, the solution is $x = $ **-2.28** to 2 decimal places.
Note: This could be quite tedious due to time constraints. A quicker method of doing this is by using a scientific calculator with the $\boxed{\text{ANS}}$ sign.

Type in the first approximation (x_1) into your calculator and press the $\boxed{= \text{sign}}$ button. This sets it up for use with the iteration formula while entering 'ANS' and pressing the '= sign' button each time to get x_3, x_4, x_5\ldots\ldots\ldots

Example 2: Work out the first five iterations using the iterative formula $x_{n+1} = 4x_n - 3$ with $x_1 = 2$.
Solution: $x_2 = 4 \times 2 - 3 = 5$, $x_3 = 4 \times 5 - 3 = 17$, $x_4 = 65$, $x_5 = 257$ and $x_6 = 1025$.

Facts about iteration

If the x value does not converge to a particular number during iteration, re-arrange the equation in another way.

The equation $x^3 - 7x + 3 = 0$ can be re-arranged in the following ways:

1) $x^3 - 7x + 3 = 0$

Make this x the subject of the formula
add (**7x**) to both sides
$x^3 + 3 = 7x$

divide by **7** on both sides
$$x = \frac{x^3 + 3}{7}$$

Using the starting number as **4**, the solution will not tend to a limit. It is **diverging** as the values are not tending to a particular number. In this case, we must re-arrange using the other unknown in the other expression of the same equation (though same letter, x).

2) $x^3 - 7x + 3 = 0$

Make this x the subject of the formula.
add (7x) to both sides
$x^3 + 3 = 7x$

Subtract 3 from both sides
$x^3 = 7x - 3$
$$x = \sqrt[3]{7x - 3}$$

Using the value of $x_1 = 4$ as the starting number, x equals 2.40 to 2 decimal places after several iterations. It tends to **converge** at x = 2.40.

Finally, if the starting number is not given, you must estimate its value and use it in the iteration formula created. Consult your teacher for assistance in using the scientific calculator as you will continue the process until the required degree of accuracy is obtained.

Care must also be taken to enter the correct **starting number** and other forms of calculator manipulations.

Using the scientific calculator for subsequent iterations is vital to avoid unnecessary time-wasting which could lead to frustrations and mistakes.

EXERCISE 3

1) Work out the first five iterations of the iterative formula $x_{n+1} = 2x_n + 3$.
Use the starting number as 3.

2) (a) Show how to rearrange the equation $x^3 + 7x - 3 = 0$ to give $x = \dfrac{3}{x^2 + 7}$

(b) If $x_1 = 5$, find a solution to the equation using iteration correct to 2 decimal places.

(c) Show that there is another way of rearranging the equation $x^3 + 7x - 3 = 0$ to give
$x = \sqrt[3]{3 - 7x}$. Show all working out.

3) A field is in the form of a rectangle. The width is 3m less than the length and the area of the field is 64m².

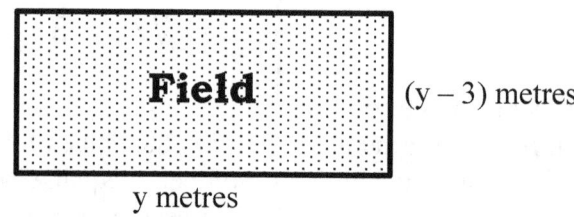

Field $(y - 3)$ metres

y metres

(a) Write an expression for the area of the field.
(b) Write a quadratic equation for the area of the field
(c) Write an iterative formula and (d) Find the length of the field y, using the iterative formula to 2 decimal places. Let the starting number = 2.

4) Look at this equation: $x^3 - 8x + 7 = 0$

(a) Show that the equation can be rearranged to give: $x = \sqrt[3]{8x - 7}$
(b) Write the iterative formula.
(c) Using **4** as the starting number find a solution using the iterative formula, correct to 2 decimal places.

5) By using a reasonable starting number, find a solution to the quadratic equation $x^2 - x - 7 = 0$ using iteration. Show all working out and correct to 3 significant figures

6) $\boxed{x_{n+1} = \dfrac{x_n^2 + 7}{3}}$ is the iterative formula used to find a solution to the quadratic
equation $x^2 - 3x + 7 = 0$.

If $x_1 = 2$ as the starting no, what happens with the iteration?

4 Quadratic Inequalities

This section covers the following topics:

- Representing quadratic inequalities
- Solving quadratic inequalities

LEARNING OBJECTIVES

By the end of this unit, you should be able to:

a) Understand and represent quadratic inequalities on a number line

b) Solve quadratic inequalities

KEYWORDS

- Quadratic
- Inequality
- Number line
- Solve

4.1 INEQUALITIES ON A NUMBER LINE

Recap: The number line can be used to show solutions to linear inequalities. The following conventions show how to represent inequalities.

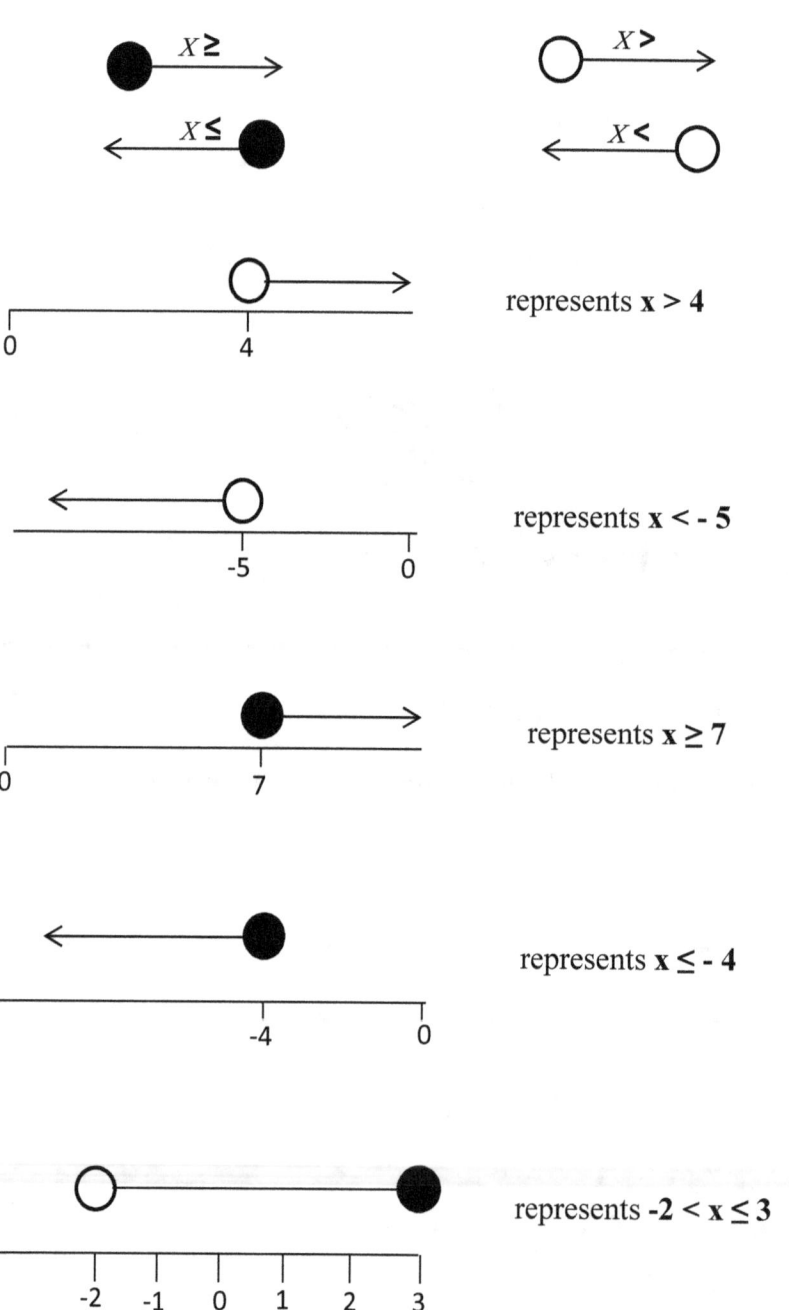

represents x > 4

represents x < - 5

represents x ≥ 7

represents x ≤ - 4

represents -2 < x ≤ 3

4.2 SOLVING QUADRATIC INEQUALITIES

When solving inequalities, imagine the inequality sign to be an **equal sign** and solve the equation. However, remember to put back the inequality sign when completing the equation. The inequality sign may change to the opposite sign if: you are dividing or multiplying by a negative number.

Example 1

Solve this quadratic inequality $x^2 < 25$ and represent the solution on a number line.

Solution:
Imagine the inequality sign ($<$) to be an equal sign.
$x^2 = 25$
$x = \pm \sqrt{25}$

The solutions are $x = +5$ and $x = -5$

Now, change the equality sign to the inequality sign ($<$).

The solution $x = 5$ would suggest the condition $x < 5$.

Therefore, **$x < 5$** satisfies the inequality $x^2 < 25$.

From the solution $x = -5$, the condition to be obtained cannot be $x < -5$ as this **does not** satisfy the inequality $x^2 < 25$. For example, if x is -6 which is less than -5, $(-6)^2$ is 36 and 36 is not less than 25.

We must now reverse the sign to **$x > -5$** as it satisfies the inequality $x^2 < 25$.
The solutions to the quadratic inequality $x^2 < 25$ are **$x < 5$** and **$x > -5$.**

To represent the solutions on a number line, you must remember that $x > -5$ can be written as $-5 < x$.

Therefore, the solution to $x^2 < 25$ could be written as **$-5 < x < 5$** and will be represented on a number line as

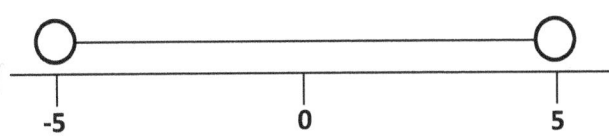

Example 2
Solve the quadratic inequality $x^2 \geq 9$ and show the solution on a number line.

Solution:
Change the inequality sign to the equality sign, (=)
$x^2 = 9$ $\Longrightarrow$ $x = \pm\sqrt{9}$

Therefore, $x = 3$ and $x = -3$

Looking at the positive value of x would suggest the condition $x \geq 3$.

Therefore, **$x \geq 3$** satisfies the inequality $x^2 \geq 9$.

From the solution $x = -3$, the condition obtained cannot be $x \geq -3$ as this **does not** satisfy the inequality $x^2 \geq 9$. The inequality sign must now be reversed to **$x \leq -3$** and this satisfies the inequality $x^2 \geq 9$.

The solutions to the quadratic inequality $x^2 \geq 9$ is **$x \geq 3$ and $x \leq -3$.**
On a number line, the solution will be represented as

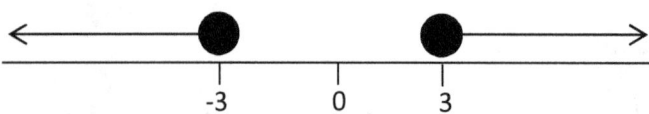

$\qquad$ -3 $\qquad$ 0 $\qquad$ 3

Example 3 Solve $x^2 - 3x - 10 < 0$

Factorise the left side to $(x + 2)(x - 5)$ $\longrightarrow$

> Find two numbers that will multiply to give -10 but will add to give -3. The numbers are -5 and 2

If $(x - 5)(x + 2) = 0$,
Either $x - 5 = 0 \Longrightarrow x = 5$
or $\quad x + 2 = 0 \Longrightarrow x = -2$

From the solution $x = 5$, the condition would suggest the solution **$x < 5$** which really satisfies the inequality $x^2 - 3x - 10 < 0$.

From the solution $x = -2$, the condition obtained cannot be $x < -2$ as this **does not** satisfy the inequality $x^2 - 3x - 10 < 0$. The sign is then reversed to **$x > -2$** which will now satisfy the inequality given.

Therefore, the solution to the quadratic inequality $x^2 - 3x - 10 < 0$ is **-2 < x < 5**

EXERCISE 4A

Solve the quadratic inequalities below **and** represent their solutions on a number line.

1) $x^2 < 36$
2) $x^2 < 4$
3) $x^2 < 100$
4) $x^2 > 36$
5) $x^2 > 4$
6) $x^2 > 100$
7) $x^2 \leq 64$
8) $x^2 \leq 49$
9) $x^2 \leq 100$
10) $x^2 \geq 64$
11) $x^2 \geq 1$
12) $x^2 \geq 100$

EXERCISE 4B

Solve the inequalities **and** show their solutions on a number line.

1) $x^2 + 9 < 25$
2) $x^2 - 10 < 26$
3) $x^2 - 10 > 54$
4) $x^2 + 7 \leq 56$
5) $x^2 + 20 < 101$
6) $x^2 - 3 \leq 13$
7) $2x^2 - 4 > 28$
8) $2x^2 + 7 \leq 207$
9) $x^2 + 2x - 35 < 0$
10) $x^2 + 5x - 6 \geq 0$
11) $x^2 - 7x - 30 \leq 0$
12) $x^2 + x - 20 > 0$
13) $x^2 + 4x < 32$

5 Functions

This section covers the following topics:

- Basic functions
- Inverse functions
- Composite functions

LEARNING OBJECTIVES

By the end of this unit, you should be able to:

a) Find the input and output of a function
b) Find the inverse function
c) Find the composite of two functions

KEYWORDS

- Input
- Output
- Inverse function
- Composite function

5.1 FUNCTIONS – INPUT AND OUTPUT

A rule for changing one number into another is referred to as a **function**.

$y = 5x + 1$ is a function. The value of x is the variable and can take any value(s).

Function notation could be used in an equation. $f(x) = 5x + 1$.

The (f) identifies the expression as a function, and the (x) is for the **input**. We may also use a **function machine** to find the **output** when an input is used in the expression or equation.

Example 1: A function machine for the function notation $f(x) = 5x + 1$ could be represented as:

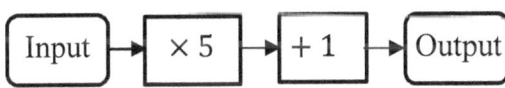

If $x = 2$, it means the input is 2.
$2 \times 5 + 1 = 11$.
It means the output is 11.

Example 2: For the function $f(x) = 3x^2 + 5x + 1$, find
a) f(4)   b) f(-5)   c) f($\sqrt{36}$)

Solutions:
a) $f(4) = 3(4)^2 + 5(4) + 1$
$= 48 + 20 + 1 = 69$
b) $f(-5) = 3(-5)^2 + 5(-5) + 1$
$= 75 - 25 + 1 = 51$
c) $f(\sqrt{36}) = 3(\sqrt{36})^2 + 5(\sqrt{36}) + 1$
$f(6) = 108 + 30 + 1 = 139$

Example 3: The function f(x) is defined as $f(x) = 2x^2 - 4$. Find the value of each of the following: a) f(-3)   b) f($\sqrt{5}$).

Solutions:
a) $f(-3) = 2(-3)^2 - 4 = 18 - 4 = 14$
b) $f(\sqrt{5}) = 2(\sqrt{5})^2 - 4 = 10 - 4 = 6$

EXERCISE 5A

1) For the function $f(x) = 3 + 7x$, find the values of:

a) f(3)   b) f(-2)   c) $f(\frac{1}{7})$ d) f($\sqrt{4}$)

2) The function machine is as shown below:

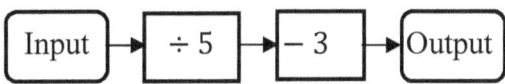

Find the output when the input is
a) 35   b) $\sqrt{100}$ c) 6   d) $\frac{1}{4}$

3) From the function machine in question 2, work out the **input** if the output is
a) 52   b) -43   c) $\sqrt{225}$ d) $\frac{3}{4}$.

4) For the function $f(x) = 2x^2 - 5$, find the values of: a) f (4)   b) f(-10)
c) f ($\sqrt[3]{8}$)   d) f($\sqrt{3}$) e) When f(x) = 27, work out **both** values of x.

5) If $g(x) = 5g^3 - 3g - 3$, find the values of: a) g(3) b) g(-7) c) g($\sqrt[4]{81}$) d) g(0).

6) Four consecutive numbers were inserted into the function $f(x) = 6x + 2$ and the sequence created was 68, 74, 80 and 86. Find the four numbers.

5.2 INVERSE FUNCTION

If a function performs the opposite process of the initial or original function, it is said to be an **inverse function.** If the original function is multiplying and the function performs division, it is an inverse function.

The notation that will be used for inverse function is $f^{-1}(x)$. However, other letters may be used.

Example 1: Find the inverse of the function $f(x) = x + 5$.

RULES FOR INVERSE FUNCTION

a) Write the function $f(x)$ as y.

b) Make x the subject of the formula or function.

c) Finally, replace x with $f^{-1}(x)$ and then replace **y** with **x**.

Solution: From $f(x) = x + 5$,

Write $f(x)$ as y.

$y = x + 5$.

Make x the subject. $y - 5 = x$

Finally, replace x with $f^{-1}(x)$ and change y to x. This will give:

$f^{-1}(x) = x - 5$ ✓

Example 2: Find the inverse function $f(x) = 6x - 3$.

Solution: Write f(x) as y $\longrightarrow$ $y = 6x - 3$.

Make x the subject of the formula.
$y + 3 = 6x$ $\longrightarrow$ $x = \frac{y+3}{6}$

$f^{-1}(x) = \frac{x+3}{6}$ ✓

Example 3: Find the inverse function $f(x) = x^2 - 7$.

Solution: $y = x^2 - 7$

$y + 7 = x^2$ $\longrightarrow$ $x = \sqrt{y+7}$

Finally, $f^{-1}(x) = \sqrt{x+7}$ ✓

Example 4: Find the inverse function

$f(x) = \frac{x+5}{2x-3}$.

Solution: $y = \frac{x+5}{2x-3}$

To make x the subject, multiply both sides by $2x - 3$.
$y(2x - 3) = x + 5$

$2xy - 3y = x + 5$

$2xy - x = 5 + 3y$

$x(2y - 1) = 5 + 3y$

$x = \frac{5+3y}{2y-1}$

Finally, $f^{-1}(x) = \frac{5+3x}{2x-1}$ ✓

EXERCISE 5B

1) Find an expression for the inverse function $f^{-1}(x)$ for:

a) $f(x) = x - 7$

b) $f(x) = 2x + 9$

c) $f(x) = x^2 - 1$

d) $f(x) = \frac{x+2}{3}$

e) $f(x) = \frac{3}{x+5}$

f) $f(x) = x^3 - 10$

2) From questions 1a - f above, find the value of $f^{-1}(8)$.

3) If $f(x) = \frac{5x-7}{2x+9}$, find an expression for $f^{-1}(x)$.

4) Find the inverse function of the following:

a) $f(x) = \frac{4}{x} + 3$

b) $f(x) = \frac{3x+3}{17}$

c) $f(x) = \frac{13}{x}$

d) $f(x) = \frac{px+q}{rx-p}$

5) Given that $f(x) = \frac{8x^2 - 22x + 5}{4x^2 - 25}$,

Find a) $f^{-1}(x)$ b) $f^{-1}(5)$

5.3 COMPOSITE FUNCTIONS

A **composite function** is when a third function is created by combining two functions.

If **f(x)** and **g(x)** are two functions, the function formed by substituting f(x) into g(x) is known as **gf(x)**. Also, the function formed by substituting g(x) into f(x) is called **fg(x)**.

Example 1: If $f(x) = 3x + 2$ and

$g(x) = x - 1$, find a) fg(x)   b) gf(x).

Solution: **a)** fg(x) means substitute g(x) into f(x). $f(x) = 3x + 2$

$$= 3(\textbf{x - 1}) + 2$$

$$= 3x - 3 + 2 = \textbf{3x - 1}$$

b) gf(x) means substitute f(x) into g(x).

$g(x) = x - 1 = (3x + 2) - 1$

$$= 3x + 2 - 1 = \textbf{3x + 1}$$

Example 2: $f(x) = \frac{2}{5}x + 5$ and

$g(x) = 2x - 1$. Find i) fg(3) and ii)gf(-2).

Solution: **a)** $g(3) = 2(3) - 1 = 5$

$$fg(3) = f(5) = \frac{2}{5}(5) + 5 = \textbf{7}$$

b) $f(-2) = \frac{2}{5}(-2) + 5 = 4.2$

$gf(-2) = g(4.2) = 2(4.2) - 1 = \textbf{7.4}$

Example 3: Given that $f(x) = 6x + 3$ and $g(x) = 4x$, find a) $fg(x)$ b) $ff(x)$.

<u>Solution:</u> **a)** $fg(x) = 6(4x) + 3 = \mathbf{24x + 3}$

b) $ff(x) = 6(6x + 3) + 3$

$$= 36x + 18 + 3 = \mathbf{36x + 21}$$

Example 4:

Given that $g(x) = 5x^2 + 6x - 4$ and

$f(x) = x + 2$, find $g(fx)$.

<u>Solution:</u> Put $f(x)$ into $g(x)$.

$gf(x) = 5(x + 2)^2 + 6(x + 2) - 4$

$$= 5(x^2 + 4x + 4) + 6x + 12 - 4$$

$$= 5x^2 + 20x + 20 + 6x + 8$$

$$= \mathbf{5x^2 + 26x + 28}$$

Example 5: $f(x) = x^3 - 3$ and

$g(x) = 2(x - 5)$. Find simplified expressions for **a)** $gf(x)$ **b)** $gg(x)$.

<u>Solution:</u>

a) $gf(x)$ means substitute $f(x)$ into $g(x)$. Expand $g(x) = 2(x - 5)$ to $2x - 10$.

$gf(x) = 2(x^3 - 3) - 10$

$$= 2x^3 - 6 - 10 = \mathbf{2x^3 - 16}$$

b) $gg(x)$ means substitute $g(x)$ into $g(x)$.
$gg(x) = 2(2x - 10) - 10 = 4x - 20 - 10$

$$= \mathbf{4x - 30}$$

EXERCISE 5C

1) Given the functions

$f(x) = 5x - 1$, $g(x) = x^2 - 3$ and

$h(x) = 9 - x^2$, determine the simplified expressions for each of the following.

a) $fg(x)$

b) $gf(x)$

c) $fh(x)$

d) $ff(x)$

2) Given the functions $f(x) = 4 - 2x$ and $g(x) = \frac{x+2}{3}$, find the value of:

a) $fg(2)$

b) $fg(-5)$

c) $gf(2)$

d) $gg(-10)$

3) If $f(x) = 6 - x$, $g(x) = 7(x - 3)$, find expressions for each of the following.

a) $gg(x)$ b) $fg(x)$ c) $gf(x)$

Using the functions above, find the value of each of the following:

d) $fg(1.2)$

e) $gf(-3)$

4) Given $f(x) = 2x^2 - 3x + 5$ and $g(x) = (x - 7)$, find $gf(x)$.

6 Circles and Equation

This section covers the following topics:

- Equation of a circle
- Tangents to a circle

LEARNING OBJECTIVES

By the end of this unit, you should be able to:

a) Find the radius and diameter of a circle
b) Find the equation of a tangent to a circle
c) Form the equation of a circle

KEYWORDS

- Circle
- Tangent
- Equation

6.1 EQUATION OF A CIRCLE

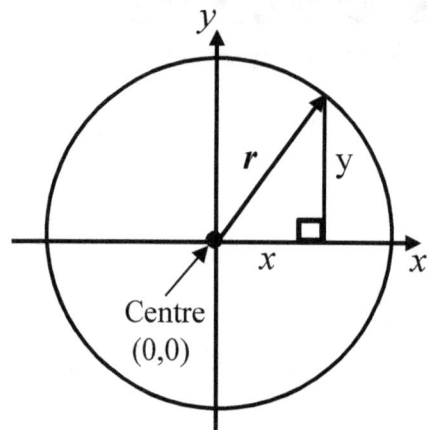

Every point on the circumference is the same distance from the centre of the circle. It is the radius.

When centre is (0,0) and radius is r, the circle has equation $x^2 + y^2 = r^2$.

Example 1: The graph of a circle is given by the equation $x^2 + y^2 = 4$.

a) Write down the radius of the circle.

b) Draw the graph of the circle.

c) What are the coordinates of the points where the graph cuts the x and y-axes?

d) What is the diameter of the circle.

Solutions: **a)** $r^2 = 16$, therefore, $r = \sqrt{4} = 2$. Radius is **2** units.

b)

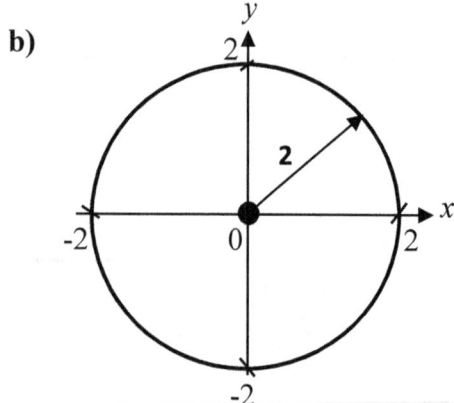

c) x – axes (-2, 0) and (2, 0)
  y – axes (0, -2) and (0, 2)
d) Diameter = $2 \times radius = 2 \times 2 = $ **4 units**.

Example 2: A circle has equation $x^2 + y^2 = 25$.

Determine if the points
a) (2, 5), b) (1, 2) and c) (-3,4) lie
outside the circle, *inside* the circle or on the *circumference* of the circle.

Solutions: Pythagoras' theorem will be used to test the points.
a) For point (2, 5):

$2^2 + 5^2 = 4 + 25 = 29$.
29 is greater than 25, therefore the point will lie **outside** the circumference.

b) For point (1, 2):
$1^2 + 2^2 = 1 + 4 = 5$.
5 is less than 25, therefore the point will lie **inside** the circle.

c) For point (-3, 4):
$(-3)^2 + 4^2 = 9 + 16 = 25$.
25 is equal to 25, therefore the point will lie on the **circumference** of the circle.

Example 3: A circle has equation $x^2 + y^2 = 25$. Point P (3, 4) is on the circumference of the circle. Work out the equation of the tangent of the circle at point P.

Solution: Sketch the circle.

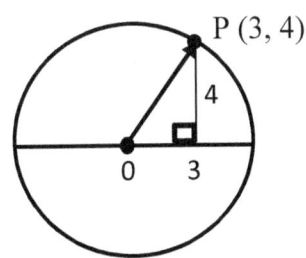

Work out the gradient of the radius.

Gradient of the radius $(m_1) = \frac{4}{3}$

Next is to find the gradient of the tangent (perpendicular) at (3, 4).

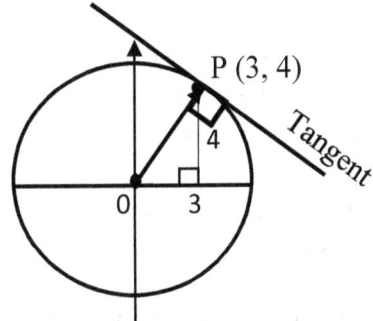

Remember Gradient of a perpendicular line $(m_2) \times$ gradient of the radius (m_1) must equal **-1**.

Therefore, $m_2 = -\frac{3}{4}$
From $y = mx + c$, and using point (3, 4),
$4 = -\frac{3}{4} \times 3 + c$
$4 = -\frac{9}{4} + c$.
Add $\frac{9}{4}$ *to both sides*
$c = \frac{25}{4} = 6.25$
Replacing c = 6.25 in the equation,
$y = mx + c$,
$y = -\frac{3}{4}x + 6.25$

Therefore, the equation of the tangent at point P is:
$y = -\frac{3}{4}x + 6.25$

EXERCISE 6

1a) Draw x and y-axes from -4 to 4.

b) Draw the graph of $x^2 + y^2 = 9$.

c) What is the diameter of the circle?

d) Write down the coordinates of the points where the circle touches the x and y-axes.

2a) Draw the graph of $x = 3$ using the axes for x and y marked from -6 to 6.

b) On the same graph, draw a circle with a diameter of 10 units, centre (0,0).

c) Work out the equation of the circle.
d) Write down the coordinates of the points where the circle touches the x and y-axes.

e) Write down the coordinates of the points where line x=3 touches the circle.

3a) Draw the graph of the circle $x^2 + y^2 - 64 = 0$.

b) On the same axes, draw the graph of $y = 3x + 2$.
c) Find the x - coordinates of the points where the graphs intersect.
d) What is the gradient of the graph $y = 3x + 2$?
e) Draw another line parallel to $y = 3x + 2$ and passing through the point, (0,6). Label it E.
f) Find the x - coordinates of the points where line E touches the circle.

4) A circle has equation $x^2 + y^2 = 80$. Point C (-4, 8) lies on the circumference of the circle.

a) Find the gradient of the tangent to the circle at point C.

b) Work out the equation of the tangent to the circle at point C.

5) The circle $x^2 + y^2 - 121$ has tangents at points (0,11) and (11,0). Work out the equations of the tangents.

6) Form the equation of a circle with centre at the origin (0,0) when the radius is

a) 3 b) 13 c) $\sqrt{5}$ d) $3\sqrt{3}$ e) 3.9

7) A circle has equation $x^2 + y^2 = 625$. Find the equation of the tangent to the circle at these points.
a) (7, 24)   b) (-7, 24)

8) Point P (0,12) is the point where two tangents meet. The angle between the two tangents is 64°. Work out the equation of the circle.

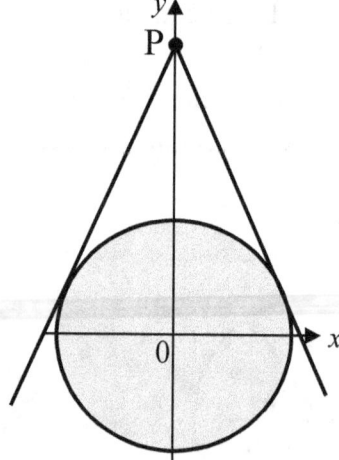

46

9)

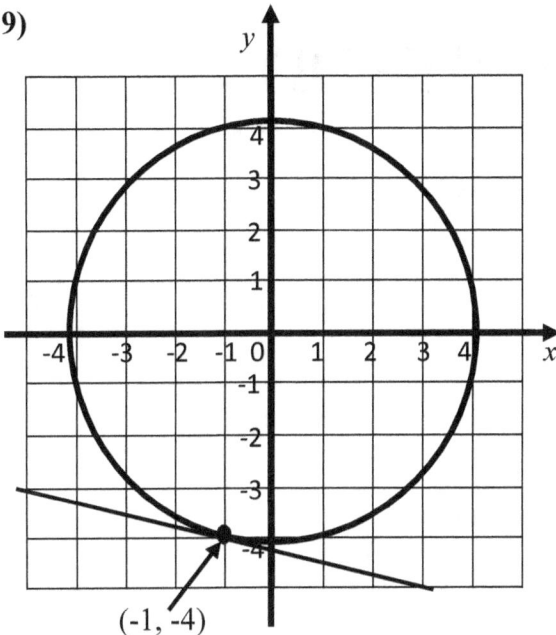

(-1, -4)

The circle above has equation
$x^2 + y^2 = 17$.

a) Work out the exact value of the diameter of the circle.

b) Find the equation of the tangent at point (-1, -4).

c) Point T (-3, $\sqrt{8}$) is on the circumference. Find the equation of the tangent at point T.

10) The equation of a circle is
$x^2 + y^2 = 3721$.

Find the coordinates of the points where

a) x = 0
b) x = 11
c) y = 60.

7 Velocity-time graphs

This section covers the following topics:

- Velocity-time graphs
- Distance from a velocity-time graph
- Work out the speed and acceleration
- The area under a curve
- Rates of change

LEARNING OBJECTIVES

By the end of this unit, you should be able to:

a) Read information from a velocity-time graph
b) Find the distance from a velocity-time graph
c) Work out/estimate area under the graph
d) Draw a tangent to estimate acceleration
e) Estimate area under a curve

KEYWORDS

- Velocity-time graph
- Acceleration
- Deceleration
- Area under the graph
- Zero gradient

7.1 VELOCITY-TIME GRAPH

Fig.24 thoroughly explains the essential knowledge and mathematics behind the velocity-time graph.

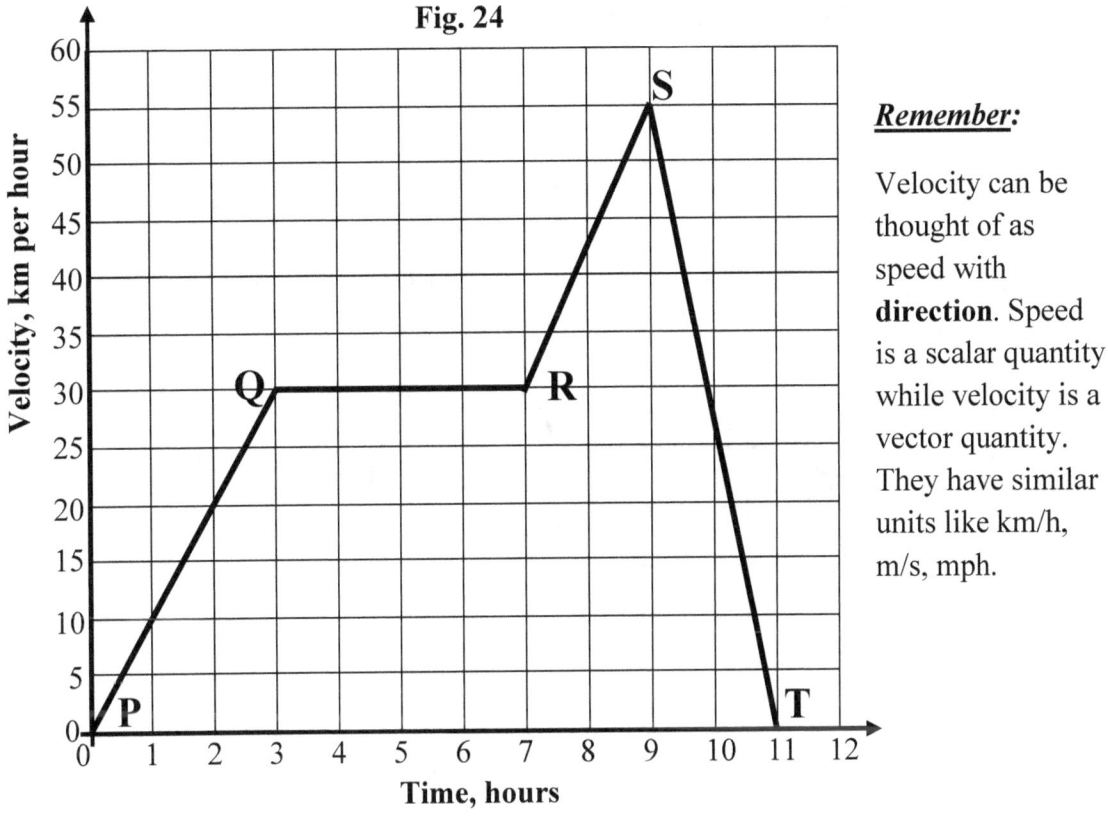

Fig. 24

Remember:

Velocity can be thought of as speed with **direction**. Speed is a scalar quantity while velocity is a vector quantity. They have similar units like km/h, m/s, mph.

Positive gradients mean the velocity is increasing. Sections PQ and RS have positive gradients. P to Q takes 3 hours and the speed increases from 0 km/h to 30 km/h. R to S takes 2 hours (9 hrs – 7 hrs) and the speed increases from 30 km/h to 55 km/h. Section QR is horizontal signifying that the speed is constant at 30 km/h for 4 hours (7 hrs – 3 hrs).

Section ST takes 2 hours as the speed decreases from 55 km/h to 0 km/h (deceleration).

For a velocity-time graph, the distance covered is the area under the graph. The total area under the graph can be calculated by splitting the graph into different sections of 2-d shapes to make it manageable.

The area under the graph will be covered at a later stage in this chapter.

Example 1: From the diagram below, calculate
a) the distance covered in the first 3 hours
b) the acceleration in the first 3 hours
c) the initial velocity
d) the total distance covered
e) the average speed for the whole journey.

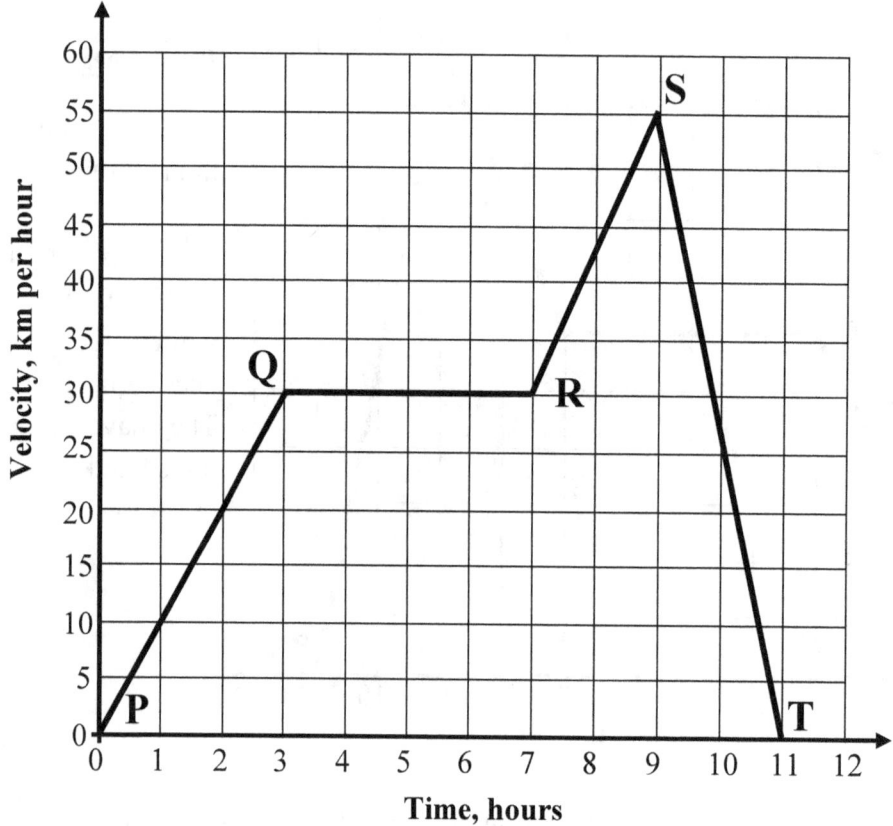

Solutions:

a) The distance in the first 3 hours = area of the triangle. $\frac{1}{2} \times 3 \times 30 = 45\ km$.

b) Acceleration in the first 3 hours = $\frac{change\ in\ veleocity}{change\ in\ time} = \frac{30-0}{3-0} = \frac{30}{3} = 10\ km/s^2$.

c) Initial velocity is the velocity at the start. This is **0 km/hour**.

d) Total distance covered is the area under the graph. Split the graph into triangles and rectangles as shown with sections A, B, C, D, E, F and G.

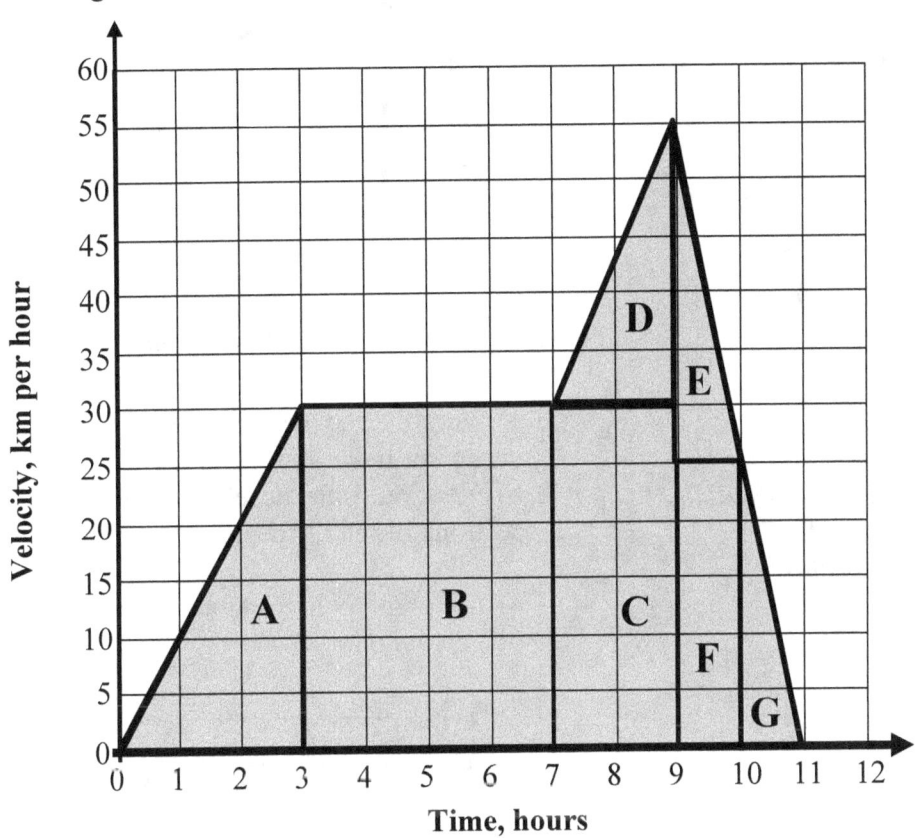

Area of triangle A $= \frac{1}{2} \times 3 \times 30 = 45$ Km^2

Area of rectangle B $= 4 \times 30 = 120$ km^2

Area of rectangle C $= 2 \times 30 = 60$ km^2

Area of triangle D $= \frac{1}{2} \times 2 \times 25 = 25$ km^2

Area of triangle E $= \frac{1}{2} \times 1 \times 30 = 15$ km^2

Area of rectangle F $= 1 \times 25 = 25$ km^2

Area of triangle G $= \frac{1}{2} \times 1 \times 25 = 12.5$ km^2

Total area under the graph
$= 45 + 120 + 60 + 25 + 15 + 25 + 12.5$
$= 302.3$ km^2

However, area under the graph is the distance travelled.

Therefore, the total distance covered
$= \textbf{302.3 km}$.

e) Average speed for the whole journey is total distance travelled ÷ total time taken.

Average speed $= \frac{302.3}{11} = \textbf{27.48 km/hour}$.

EXERCISE 7A

1) The velocity-time graph below represents a bus journey between two bus stops.

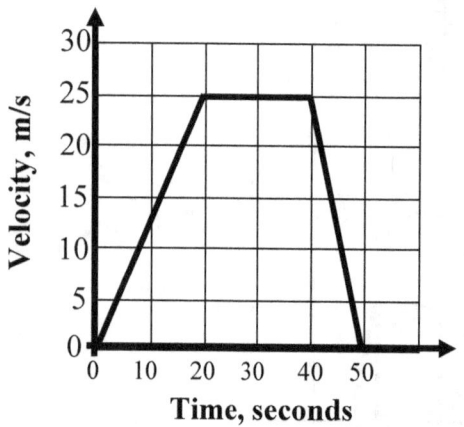

a) What is the velocity at 10 seconds?
b) What is the steady speed of the bus?
c) Find the distance between the two bus stops.
d) While decelerating, what was the distance covered by the bus?
e) Work out the average speed for the whole journey.

2)

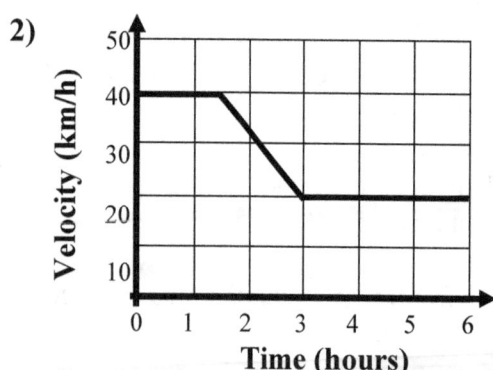

Work out the distance travelled in the first 5 hours.

3) This is a velocity-time graph.

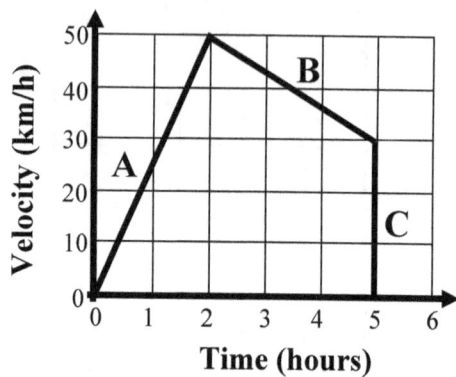

a) Work out the average speed of the entire journey.
b) Work out i) the acceleration for sections A and ii) the deceleration for B.

4) The graph below shows the velocity of a sports bike.

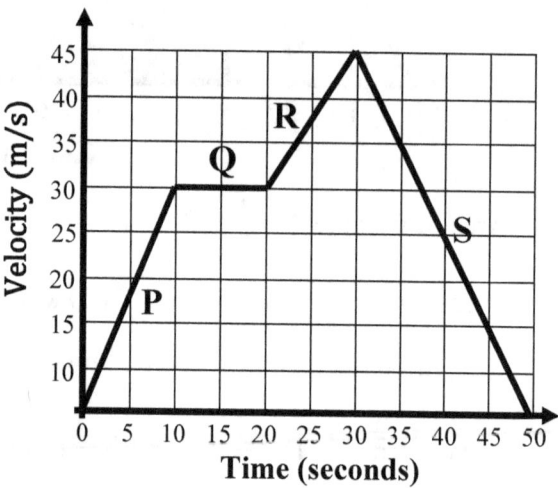

a) What is the velocity of the bike after 25 seconds?
b) Describe the movement(s) of this sports bike during P, Q, R, S sections.
c) Describe what is happening after 30 seconds.
d) At what rate is the bike decelerating?

7.2 AREA UNDER A CURVE

To work out the area under a curve accurately is beyond GCSE and would be dealt with in Advanced level.

However, we estimate the area under a curve by splitting (dividing) into smaller shapes of rectangles, triangles or trapezium.

Also, it could be **under-estimating** or **over-estimating** depending on the divisions made on the curve.

Example 1: Calculate the estimate of the total distance travelled.

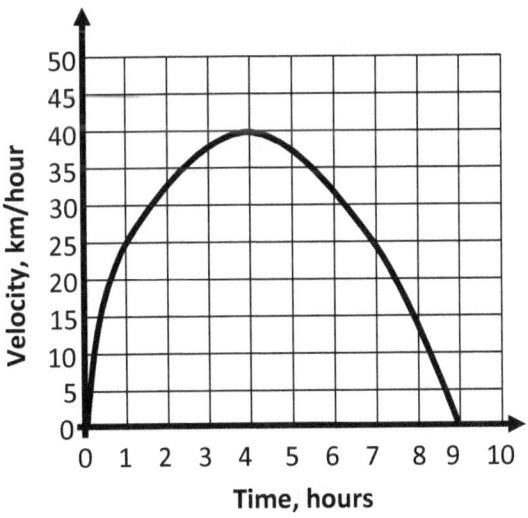

Solution: Split the shape into A, B, C and D as shown in the diagram. This is just a guide. You may split into manageable parts to help in your calculations.

Remember: Area under the graph is the **distance** travelled. Therefore, the unit will be in **km**, to represent the distance.

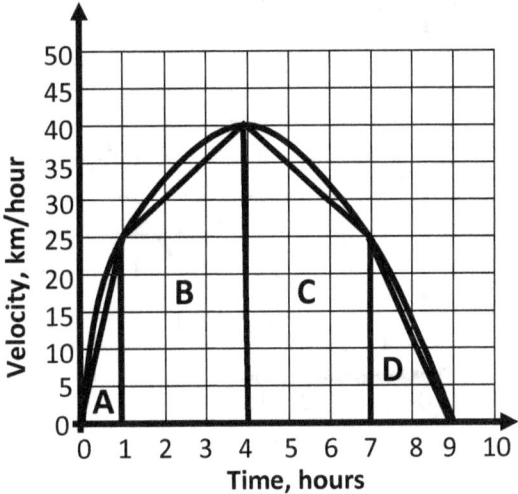

Area of triangle A $= \frac{1}{2} \times 1 \times 25$
$= 12.5$ km

Area of trapezium B $= \frac{1}{2}(25 + 40) \times 3$
$= 97.5$ km

Area of trapezium C $= \frac{1}{2}(40 + 25) \times 3$
$= 97.5$ km

Area of triangle D $= \frac{1}{2} \times 2 \times 25$
$= 25$ km

Total area $= 12.5 + 97.5 + 97.5 + 25$
$= 232.5$ km

Therefore, the total distance travelled is **232.5 km**. Since all the split shapes are inside the graph (curved area), this will represent a slight **under-estimation** of the real distance travelled.

Example 2: Estimate the distance covered.

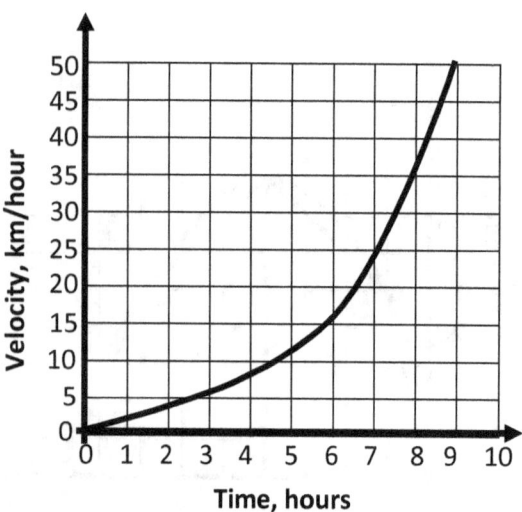

Velocity, km/hour (vertical axis), Time, hours (horizontal axis)

Solution: Split the shape into A and B.

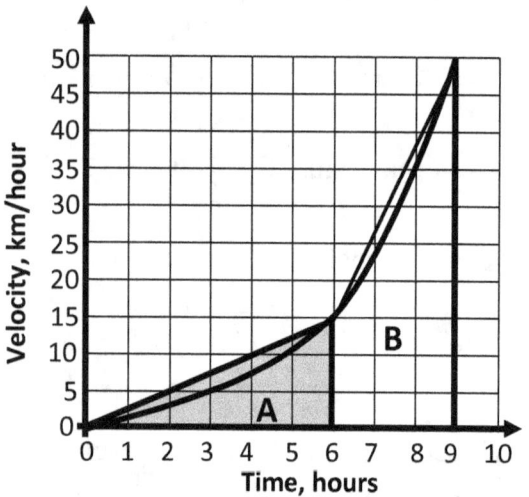

Area of triangle $A = \frac{1}{2} \times 6 \times 15 = 45$ km

Area of trapezium $B = \frac{1}{2}(15 + 50) \times 3$
$= 97.5$ km

Total distance travelled $= 45 + 97.5$
$= \mathbf{142.5}$ **km.** The areas calculated are greater than the area under the graph. Therefore, it is an **over-estimation**.

EXERCISE 7B

1) For the velocity-time graph below, estimate the distance travelled. Also, state if it was under or over-estimation.

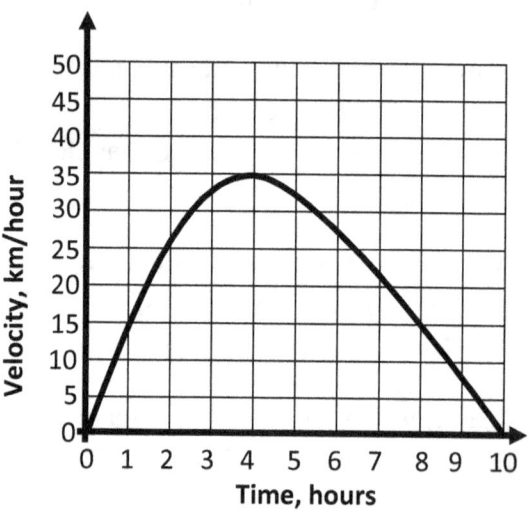

Velocity, km/hour (vertical axis), Time, hours (horizontal axis)

2) This is a velocity-time graph.

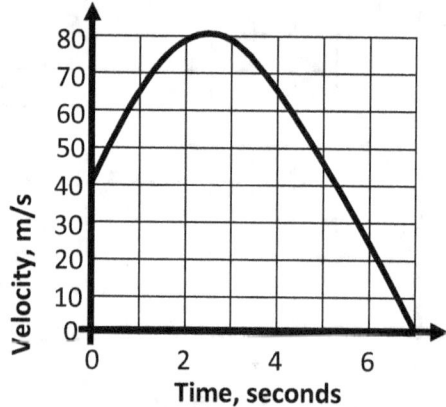

Velocity, m/s (vertical axis), Time, seconds (horizontal axis)

a) Find the initial velocity.

b) Find the maximum velocity.

c) Work out an estimate of the total distance travelled.

54

3) a) Estimate the total distance covered. b) State if it was under or over-estimation.

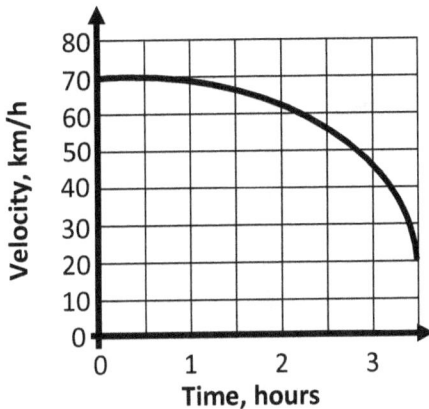

4) a) Estimate the distance covered. b) State whether it was under or over-estimation.

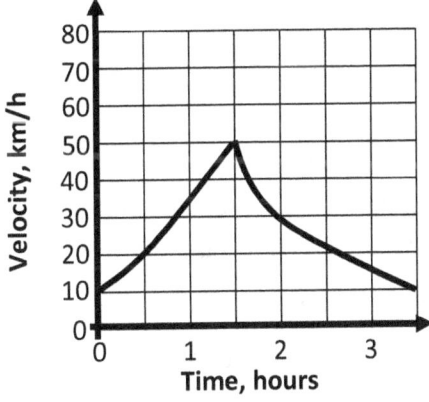

5) Estimate the total distance travelled.

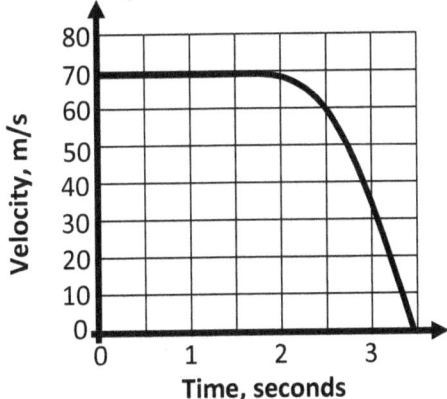

7.3 TANGENTS, GRADIENTS AND RATE OF CHANGE

This section reinforces the use of the gradient of a straight line. We shall use this knowledge to calculate acceleration at a time in a velocity-time graph.

Recall: A tangent is a straight line that touches the outside of a circle or a curve at a point only.

To accomplish this, we shall draw a tangent at the point of interest in the curve and calculate the gradient of the tangent. Draw a right-angled triangle from the tangent and calculate the $\frac{change\ in\ y}{change\ in\ x}$. That will give the acceleration at that point in time.

DISTANCE-TIME GRAPH

In a distance-time graph, the gradient at any point gives the **velocity** at that point on the graph. **NOTE**: Velocity (vector quantity) and Speed (scalar) have the same unit(s).

Example 1: Estimate the velocity at 1 hour.

Solution: Draw a tangent at 1 hour

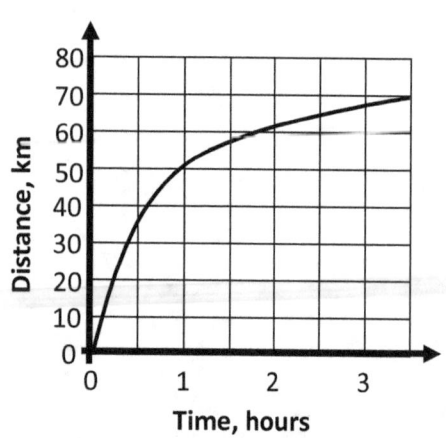

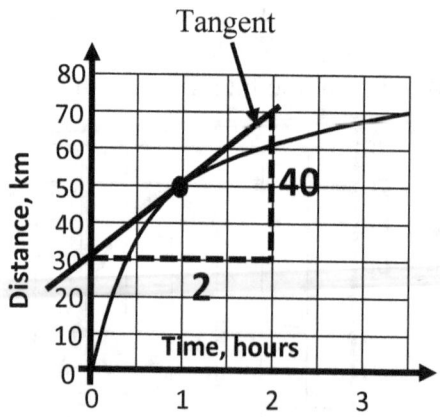

At 1 hour, velocity $= \frac{40}{2} = $ **20 km/h**

56

Average speed/velocity between two times can also be calculated by working out the gradient of the chord (line joining two points on the curve).

VELOCITY-TIME GRAPH

In a velocity-time graph, the gradient at any point gives the **acceleration** at that point on the curve.

Example 2: Estimate the acceleration at 2 hours.

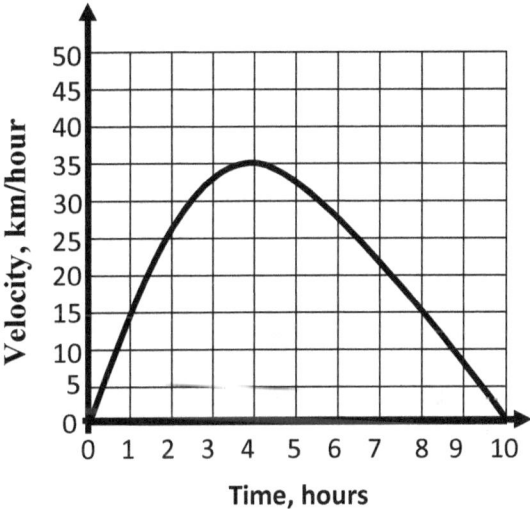

Solution: Draw a tangent at 2 hours.

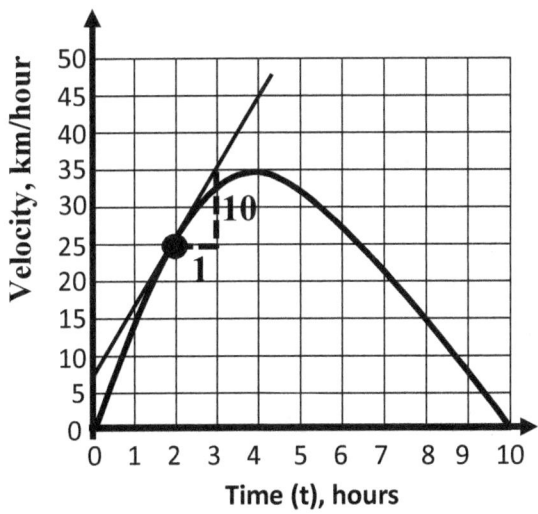

The acceleration at 2 hours is
$\frac{10}{1} = 10$ **km/h²**

Notice the unit of acceleration. In this case km/h².

From example 2, acceleration **is zero** at 4 hours. This is the highest or maximum point on the graph before deceleration starts. The gradient is zero at that point.

EXERCISE 7C

1) A distance -time graph is shown below.

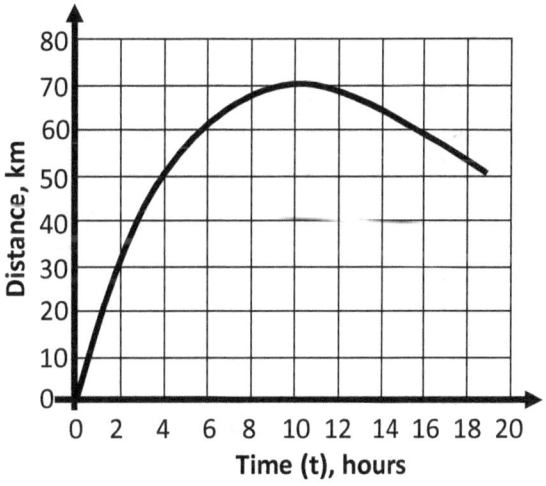

a) At what time is the velocity zero? Explain fully.

b) Estimate the velocity when:

i) t = 2 hours   ii) t = 6 hours

c) Estimate the average velocity from:

i) t = 0 hours to t = 4 hours

57

2) The graph below is a velocity-time graph.

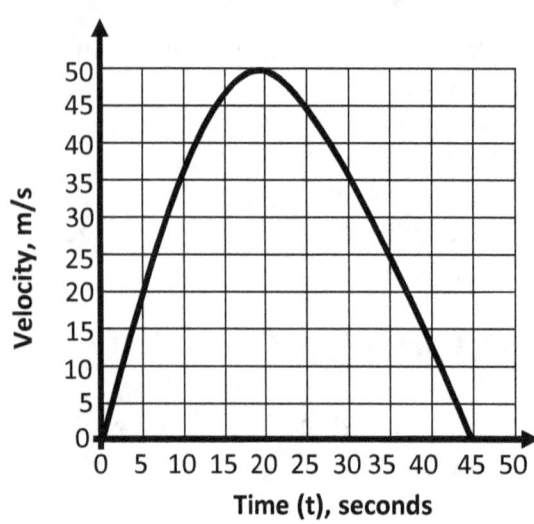

Velocity, m/s

Time (t), seconds

a) Estimate the acceleration when t = 10 s.

b) Estimate the deceleration when t = 30 s.

c) At what time is the acceleration zero? Explain fully.

3)

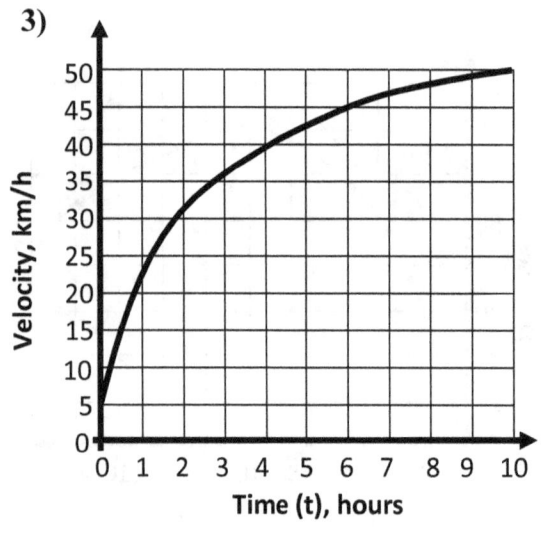

Velocity, km/h

Time (t), hours

a) Estimate the acceleration when t = 2 hrs.

b) Estimate the total distance travelled.

4)

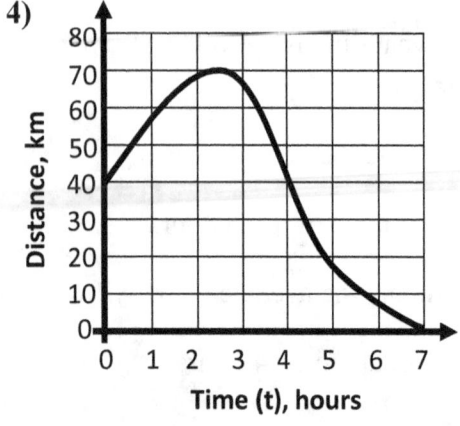

Distance, km

Time (t), hours

From the distance-time graph, estimate the velocity when:

a) t = 1.5 hrs

b) t = 2.5 hrs.

c) Estimate the average velocity from

t = 4 hours to t = 7 hours.

5) Consider the function $d = 6t - t^2$ which represents the distance, in metres, travelled by a football with respect to time of landing, t in seconds.

a) Draw the graph of the function for **t** values from 0 to 6.

b) At what time is the gradient zero?

c) What is the maximum speed?

d) At point (4,8), is the gradient positive or negative?

6) The diagram below represents a velocity-time graph of a bus moving between two counties.

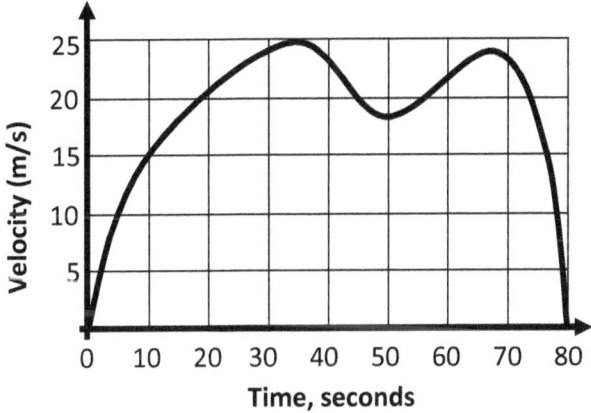

a) Estimate the acceleration of the bus when the time is equal to 10 seconds.

b) What is the acceleration when the time is equal to 35 seconds.

c) estimate the total distance travelled by the bus.

7) Consider the function $d = 8t - t^2$ which represents the distance travelled in metres by a football with respect to time of landing in seconds, t.
a) Draw the graph of the function for t values from 0 to 8.
b) Calculate the maximum speed.
c) Work out the speed when: i) t = 2   ii) t = 4.

POINTS TO NOTE:

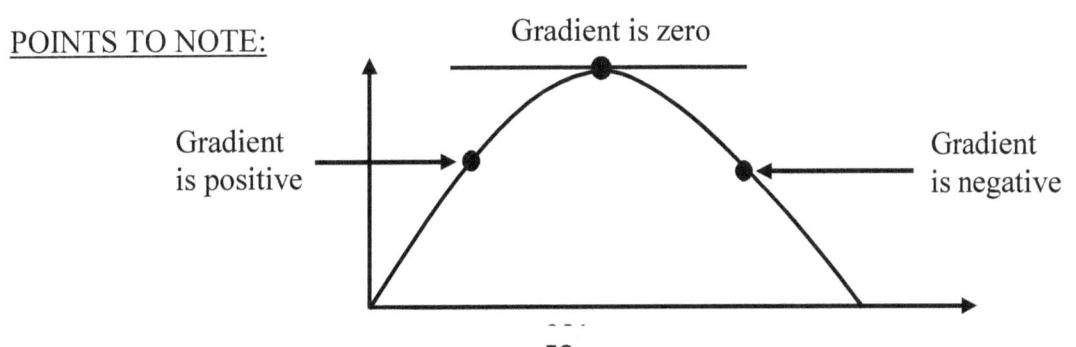

8 Exact Trigonometric Ratio

This section covers exact trigonometric ratios.

LEARNING OBJECTIVES

By the end of this topic, you should be able to:

- Work out the exact values of

  - Sin 0°, 30°, 45°, 60° and 90°

  - Cos 0°, 30°, 45°, 60° and 90°

  - Tan 0°, 30°, 45°, 60° and 90°

- Solve problems using exact trigonometric values

KEYWORDS

- Sin

- Cos

- Tan

- Exact values

- Trigonometric ratios

8.1 EXACT TRIGONOMETRIC VALUES

In this section, we shall consider working out the exact values of sin, cos and tan ratios for 0°, 30°, 45°, 60° as well as the values of sin and cos for 90°.

> **Note:** tan 90° is not defined and therefore, there is no value for it. The simple reason being that
>
> $\tan \theta = \dfrac{\sin \theta}{\cos \theta}$ and if $\theta = 90°$,
>
> $\tan 90° = \dfrac{\sin 90°}{\cos 90°} = \dfrac{1}{0}$. This is undefined and **not** allowed in maths when the denominator is zero. Hence, there is no tan ratio for 90°.

Pupils are expected to **memorise** the values of these ratios. However, some will forget and hence the reason for this section.

EXACT VALUES

Angle θ	Sin θ	Cos θ	Tan θ
0°	0	1	0
30°	$\dfrac{1}{2}$	$\dfrac{\sqrt{3}}{2}$	$\dfrac{1}{\sqrt{3}}$
45°	$\dfrac{1}{\sqrt{2}}$	$\dfrac{1}{\sqrt{2}}$	1
60°	$\dfrac{\sqrt{3}}{2}$	$\dfrac{1}{2}$	$\sqrt{3}$
90°	1	0	**Undefined**

WORKING OUT SINE, COSINE AND TANGENT RATIOS FOR 30° and 60°

An equilateral triangle of side 2 cm is drawn below. ∠ABC is bisected by BD to give 30° each.

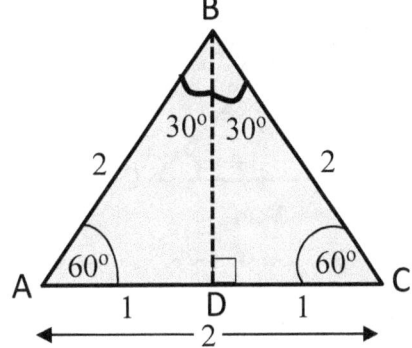

Considering triangle ABD, the length BD can be calculated using Pythagoras' theorem.

$$BD^2 = 2^2 - 1^2 = 4 - 1 = 3$$

$$BD = \sqrt{3}$$

Using SOH CAH TOA.

$$\sin 30° = \frac{opposite}{hypotenuse} = \frac{1}{2}$$

$$\sin 60° = \frac{opposite}{hypotenuse} = \frac{\sqrt{3}}{2}$$

$$\cos 30° = \frac{adjacent}{hypotenuse} = \frac{\sqrt{3}}{2}$$

$$\cos 60° = \frac{opposite}{hypotenuse} = \frac{1}{2}$$

$$\tan 30° = \frac{opposite}{adjacent} = \frac{1}{\sqrt{3}}$$

$$\tan 60° = \frac{opposite}{adjacent} = \frac{\sqrt{3}}{1} = \sqrt{3}$$

WORKING OUT SINE, COSINE AND TANGENT RATIOS FOR 45°

To show these trigonometric ratios, draw a right-angled isosceles triangle with the equal sides of 1 cm.

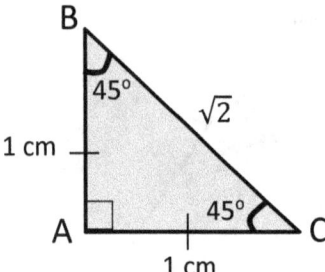

Using Pythagoras' theorem,

$$BC^2 = 1^2 + 1^2 = 1 + 1 = 2$$

$$BC = \sqrt{2}$$

Using SOH CAH TOA,

$$\sin 45° = \frac{opposite}{hyopotenuse} = \frac{1}{\sqrt{2}}$$

$$\cos 45° = \frac{adjacent}{hypotenuse} = \frac{1}{\sqrt{2}}$$

$$\tan 45° = \frac{opposite}{adjacent} = \frac{1}{1} = 1$$

Example 1: Work out the exact value of sin 60° + cos 60° without using a calculator.

Solution: $\sin 60° = \frac{\sqrt{3}}{2}$ and $\cos 60° = \frac{1}{2}$

$$\sin 60° + \cos 60° = \frac{\sqrt{3}}{2} + \frac{1}{2} = \frac{\sqrt{3}+1}{2} ✓$$

EXERCISE 8

1) Without using a calculator, calculate the **exact values** of the following.

a) sin 30° + sin 60°

b) 2 tan 30°

c) cos 30° + tan 60°

d) sin 30° + cos 60° + sin 45°

e) 4 tan 60° + cos 30°

f) tan 60° ÷ tan 30°

g) Prove that $\sin 60° + \tan 30° = \frac{5\sqrt{3}}{6}$.

2) Work out the **exact values** of the missing lengths in metres.

a)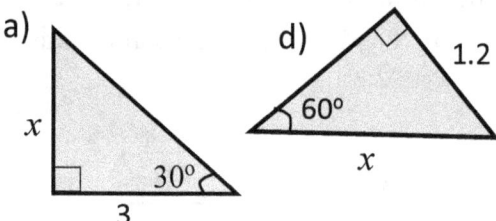

b)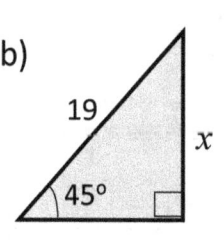

c)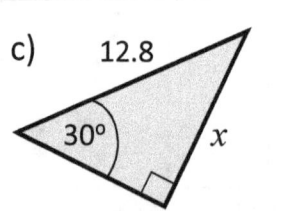

d)

e)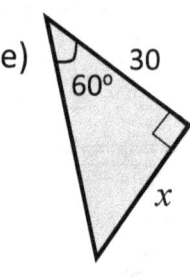

f)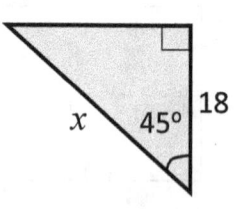

9 Venn Diagrams

This section covers the following topics:

- Sets
- Notations
- Venn diagrams and probabilities

LEARNING OBJECTIVES

By the end of this topic, you should be able to:

- Understand sets and notations
- Draw Venn diagrams
- Understand Unions and Intersections
- Understand the complement of a set
- Solve Venn diagram problems
- Work out probabilities from Venn diagrams

KEYWORDS

- Sets
- Notations
- Venn Diagram
- Probability
- Complement
- Union and Intersection

9.1 VENN DIAGRAM

In this section, we shall learn how to draw a Venn diagram, write notations using special symbols and calculate probabilities from Venn diagrams.

SETS

A **set** is a collection of numbers or objects.

If we want to represent a set of prime numbers less than 15 in a set represented by P, it should look like this:

P = {2, 3, 5, 7, 11, 13, }

P represents the set of prime numbers less than 15. Note that capital letters are mostly used to represent a set. In the set above, 2, 3, 5, 7, 11 and 13 are all **elements** within the set.

The **universal set** is the set containing **all** the elements. It is represented as ε.
We could also have a set of no element(s). It is called the **empty set, Ø**.

NOTATIONS

There are basically three special notations used in Venn diagrams.

1) A ∩ B
This means the intersection of A and B. It is the overlapping part of the Venn diagram.

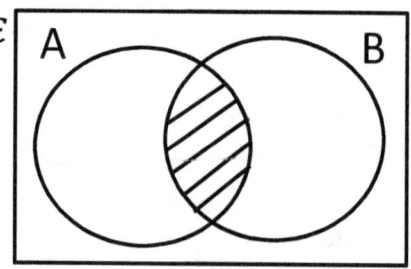

2) A ∪ B
This means the union of A and B. It is everything in A and B and is represented by the shaded part in the diagram.

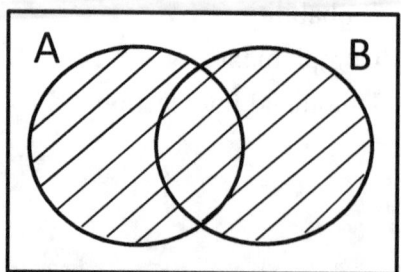

64

3) A′
This is the complement of A.
It is everything **not** in A. It is the
Shaded part.

437

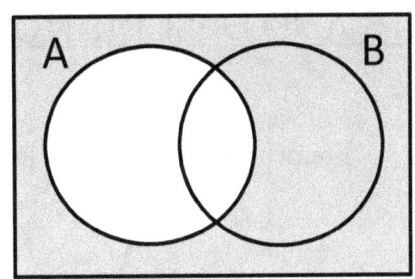

Example 1: $\varepsilon = \{2, 4, 6, 8, 10, 12, 14, 16\}$, A = $\{multiples\ of\ 4\}$ and B = $\{2, 5, 8, \}$

a) Draw a Venn diagram to illustrate the information.

b) Work out i) A $\cup$ B   ii) A $\cap$ B   iii)) A′ $\cup$ B   iv) (A $\cup$ B)′   v) A′

Solutions:

a) Venn diagram

$\varepsilon = \{2, 4, 5, 6, 8, 10, 12, 14, 16\}$
A = $\{Multiples\ of\ 4\} = \{4, 8, 12, 16\}$
B = $\{2, 5, 8, \}$

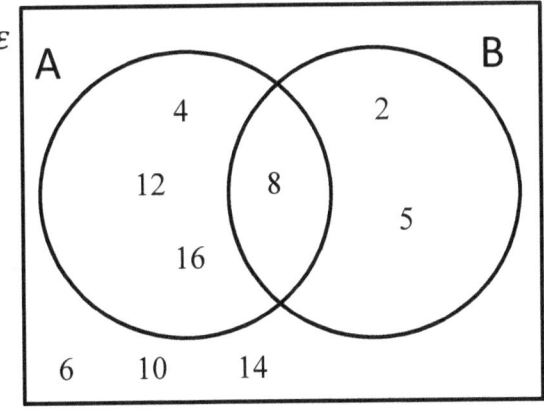

b) i) A $\cup$ B = $\{\mathbf{2, 4, 5, 8, 12, 16}\}$
These are all numbers in both A and B.

ii) A $\cap$ B = $\{\mathbf{8}\}$
This is the number in the intersection part.

iii) A′ $\cup$ B = $\{\mathbf{2, 5, 6, 8, 10, 14}\}$
These are all numbers excluding 4, 12 and 16.

iv) (A $\cup$ B)′ = $\{\mathbf{6, 10, 14}\}$
This is the complement of A Union B.
These are all the numbers **not** in A and B.

v) A′ = $\{\mathbf{2, 5, 6, 10, 14}\}$
This is the complement of A. They are all the numbers **not** in A.

PARTS AND DESCRIPTION OF THE VENN DIAGRAM

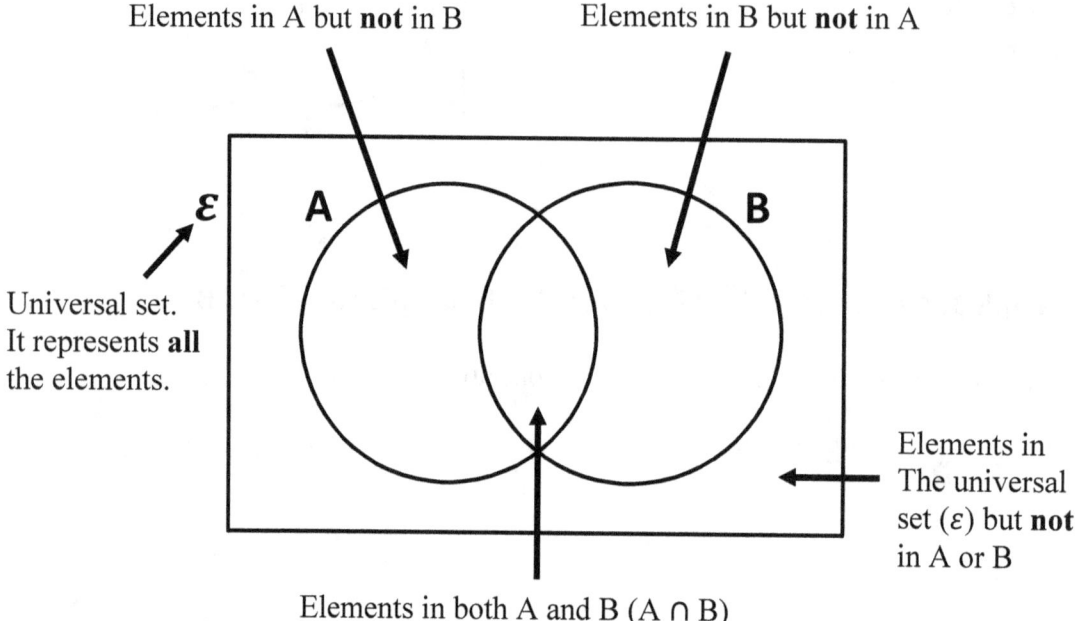

Elements in A but **not** in B

Elements in B but **not** in A

ε

A

B

Universal set.
It represents **all**
the elements.

Elements in
The universal
set (ε) but **not**
in A or B

Elements in both A and B (A ∩ B)

9.2 PROBABILITY AND VENN DIAGRAM

We may use the Venn diagram to calculate probabilities as follows.

$$P(A) = \frac{Number\ of\ elements\ in\ set\ A}{Total\ number\ of\ elements\ in\ \varepsilon}$$

Probability of A

Also:

P (A ∩ B) = P (A *and* B)

P (A ∪ B) = P (A *or* B)

P (A′) = P (*not* A)

Example 2: The Venn diagram shows the number of pupils that are right-handed (R) and the number of pupils that are left-handed (L) in a class.

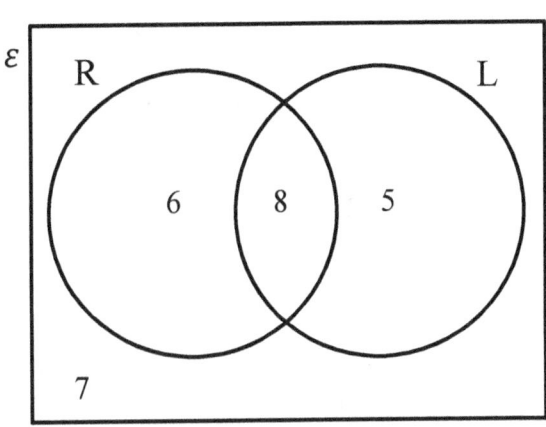

a) What is the total number of pupils in the class?

b) What is the probability that a pupil chosen at random is right handed?

Work out
c) P (R ∩ L)
d) P (R ∪ L)
e) P (R ∪ L)'
f) P (R')
g) P (R' ∪ L)
h) P (R ∩ L)'

Solution:

a) Total number of pupils = 6 + 8 + 5 + 7 = **26 pupils.**

b) 6 + 8 = 14…………14 pupils are right handed therefore, P (R) = $\frac{14}{26}$

c) P (R ∩ L) is worked out by looking at the intersection part of the Venn diagram.

$$P (R \cap L) = \frac{8}{26}$$

d) 6 + 8 + 5 = 19. Therefore, P (R ∪ L) = $\frac{19}{26}$

e) P (R ∪ L)' means the probability that the pupils are neither right nor left-handed.
This is also 1 − P (R ∪ L) = 1 - $\frac{19}{26}$ = $\frac{7}{26}$

f) R' means not in R or not right handed. This is 5 + 7 = 12. Therefore, P (R') = $\frac{12}{26}$

g) 8 + 5 + 7 = 20. Therefore, P (R' ∪ L) = $\frac{20}{26}$

h) (R ∩ L)' means everything that **is not** in the intersection part of the Venn diagram.
Everything that is not 8 is 6 + 5 + 7 = 18. Therefore, P (R ∩ L)' = $\frac{18}{26}$

67

EXERCISE 9A

1) List all the elements of the following sets.

a) P – the first six multiples of 7
b) Q – factors of 48
c) R – prime numbers between 6 and 24
d) S – double-digit cube numbers

2) The Venn diagram shows information about pupils playing flute and guitar at a music festival.

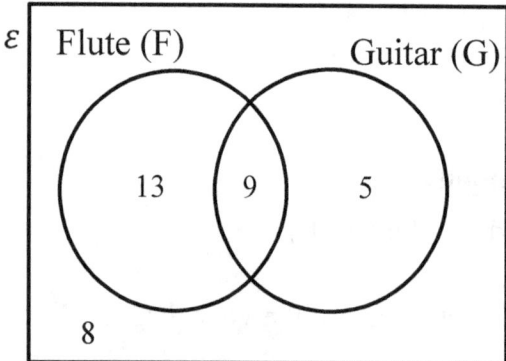

a) How many pupils were present at the festival?
b) How many pupils played the flute?
c) How many pupils played the guitar?
d) How many pupils did not play flute?
e) How many pupils played the guitar but not the flute?

If a pupil is chosen at random, find

f) P (F ∩ G)
g) P (F)
h) P (F ∪ G)
i) P (F ∪ G)'
j) P (F' ∪ G)
k) P (F ∩ G)'

3) $\varepsilon = \{1, 2, 3, 4, 5, 6, 7, 8, 9, 10, 11, 12, 13\}$
$A = \{2, 3, 7, 9\}$
$B = \{1, 5, 7, 9, 13\}$

a) Represent the information in a Venn diagram.
b) Use the Venn diagram to find:
i) P (B)
ii) P (A')
iii) P (A ∩ B)
iv) P (A ∪ B)'

4) If

$\varepsilon = \{x : x \text{ is a positive integer}, x \leq 19\}$,
list all the elements of:

a) B = $\{x : x \text{ is even}\}$
b) C = $\{x : x \text{ is a multiple of } 4\}$
c) D = $\{x : x \text{ is a prime number}\}$

5) Using set notations, describe the shaded area in each Venn diagram.

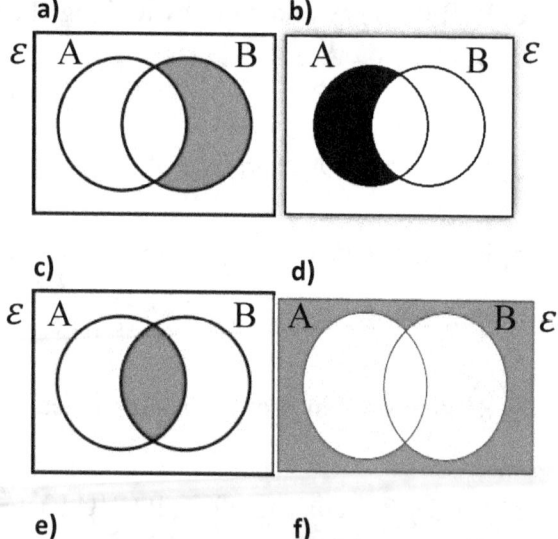

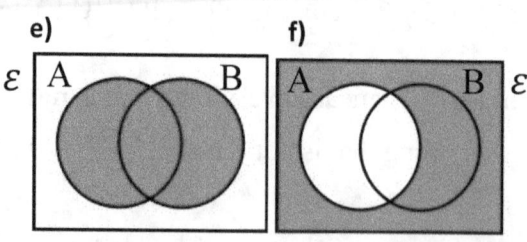

Example 3: There are 20 pupils in a class. They are each asked whether they have a dog (D) or a hamster (H).
15 say they have a dog.
8 say they have a hamster. Two pupils have neither of the pets.
Draw a Venn diagram to represent the information.

Solution:15 + 8 = 23
20 − 2 = 18
23 − 18 = 5 (overlap)

15 − 5 = 10 pupils   8 − 5 = 3 pupils

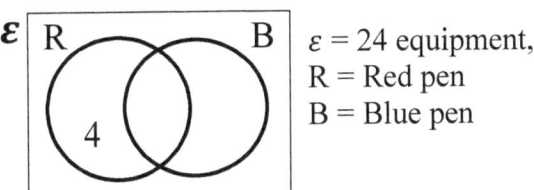

Have neither Cat nor Hamster

Example 4:
Complete the Venn diagram for pupils with equipment during an examination.

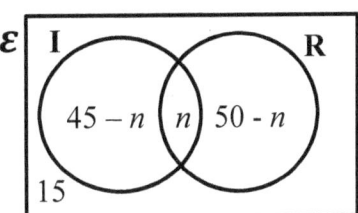 ε = 24 equipment, R = Red pen B = Blue pen

16 have red pens, 14 have blue pens, 4 have red pens but not blue pens.

Solution:
From 16 with red pens, 16 − 4 = 12
From 14 with blue pens, 14 − 12 = 2

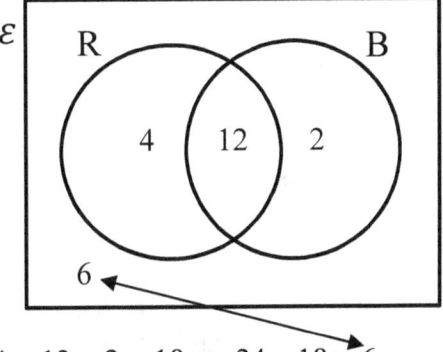

4 + 12 + 2 = 18 so, 24 − 18 = 6

Example 5: An employer has **80** workers. 45 workers are in ICT (I) department. 50 workers are right-handed (R). 15 workers are not right handed and not in ICT department.
a) Draw a Venn diagram
b) Find the number of workers who are right-handed and in ICT department.
c) Work out the probability that a worker is right handed and in ICT department.

Solution: Let the number of workers who are right-handed and in ICT department be **n**. ε = 80

ε | I   R
45 − n (n) 50 - n
15

$45 - n + n + 50 - n + 15 = 80$
$110 - n = 80$
$110 - 80 = n$
n = 30

69

a) Substituting the value of n in the Venn diagram gives

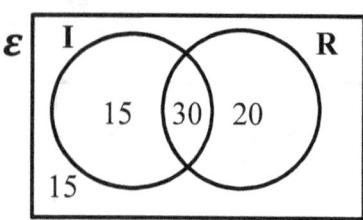

b) $n = 30$ workers

c) P (I and R) $= \frac{30}{80}$ or 0.375

Example 6: In a sixth-form class, 7 study both maths (M) and physics (P). 14 study physics but not maths and 5 study neither subject. The overall class has 32 students.
a) Draw a Venn diagram to represent the information above.
b) How many students study maths.
c) Find the probability that a student studies physics.

Solution: Let x represent the number of students who study maths only. $\varepsilon = 32$.

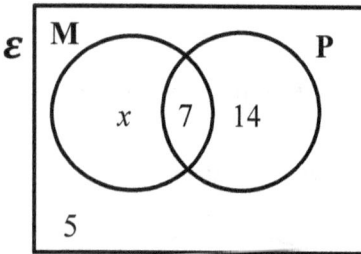

$x + 7 + 14 + 5 = 32$, so $x = 6$

a)

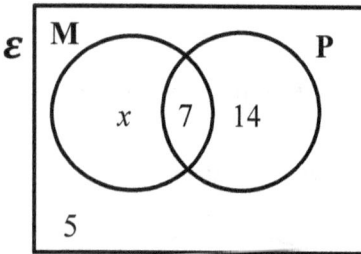

b) $6 + 7 = 13$ students

c) $7 + 14 = 21$

$P(P) = \frac{21}{32}$

Example 7: There are 25 pupils in a class. They are each asked whether they have a dog (D) or a hamster (H).
18 say they have a dog.
10 say they have a hamster. 3 pupils have neither of the pets.
a) Draw the Venn diagram.
b) Find the probability that a pupil chosen at random from the class owns a dog **given that** they own a hamster.

Solution: $18 + 10 = 28$
$25 - 3 = 22$
$28 - 22 = 6$ (overlap)
Dog only: $18 - 6 = 12$
Hamster only: $10 - 6 = 4$

a)

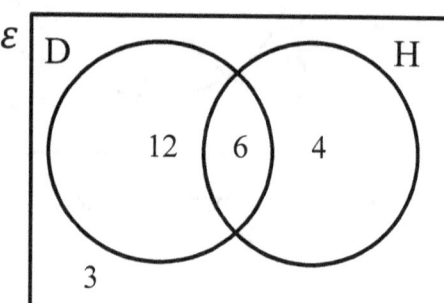

b)

$6 + 4 = 10$

Therefore, the probability $= \frac{6}{10}$

Example 8: 92 students in a sixth form college.

40 study Biology (B)
50 study Physics (P)
25 study English (E)

10 study both Biology and Physics
12 study both Physics and English
8 study both Biology and English

All the pupils study one or more of these subjects.
a) Draw a Venn diagram to illustrate the information.
b) How many pupils study all three subjects?
c) What is the probability that a pupil picked at random studies all subjects mentioned?

<u>Solution</u>: Sketch the Venn diagram. Let the number of students that studies all three subjects be x.

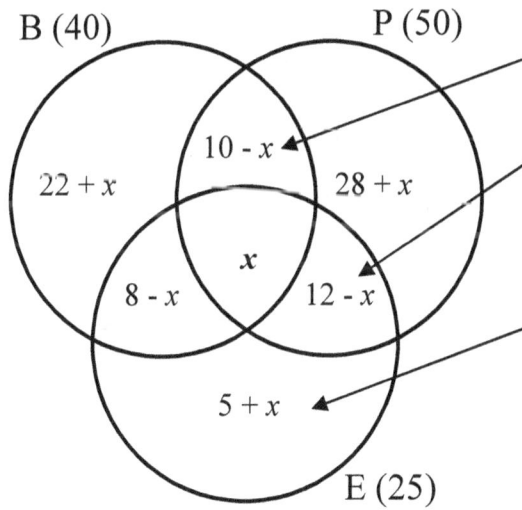

10 students study Biology and Physics. Therefore, this portion is $10 - x$

12 students study Physics and English. Therefore, this portion is $12 - x$
Do the same to the remaining part to get $8 - x$

This area $= 25 - (8 - x + x + 12 - x)$
$\qquad = 25 - (20 - x)$
$\qquad = 5 + x$
Do the same for other areas to get $22 + x$ and $28 + x$ respectively

a)

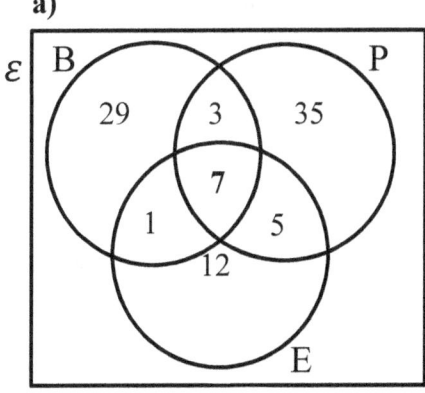

b) Add all the numbers and letters in the Venn diagrams. This will equal 92.
$22 + x + 10 - x + x + 8 - x + 28 + x + 12 - x + 5 + x = 92$. This reduces to $85 + x = 92$.
$x = 92 - 85 = 7$.

Therefore, seven **students** study all three subjects.　**c)** $\frac{7}{92}$

71

EXERCISE 9B

1) In a class of 27 pupils, 20 have a cat, 9 have a dog. 6 pupils have neither.

a) Draw a Venn diagram to represent the above information.
b) Calculate the probability that a pupil chosen at random from the class owns
i) a cat and a dog ii) a dog
c) Find the probability that a pupil chosen at random from the class owns a dog, given that they own a cat.

2) P(X) = 0.35 and P (Y) = 0.12.
Write down a) P (X′) b) P (Y′)

3) An employer has **110** workers. 65 workers are in the media (M) department. 60 workers are left-handed (L). 20 workers are not left-handed and not in the media department.
a) Draw a Venn diagram
b) Find the number of workers who are left-handed and in media department.
c) Work out the probability that a worker is left-handed and not in the media department.

4) The number of elements in the complement of B = n (B′) = 27.

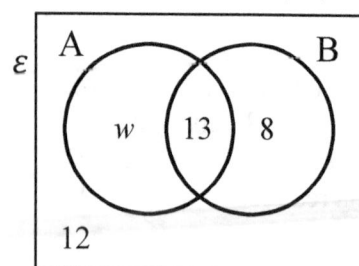

Work out
a) w b) n (A ∪ B)   c) n(A ∪ B′).

5) The Venn diagram below shows some probabilities.

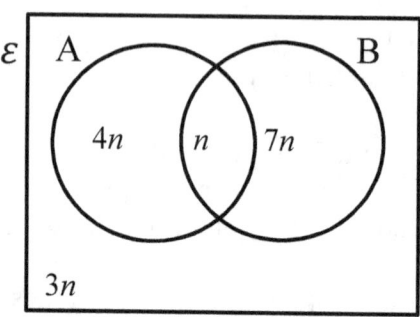

Calculate:
a) P (A) b) P (B) c) P (A′) d) P (B′)

e) P (A ∪ B) f) P (A ∩ B′) g) P (A′ ∩ B)

6) 120 pupils in a school are in year 11.

60 study Geography
55 study History
45 study French

15 study both Geography and History
18 study both History and French
12 study both Geography and French

All the pupils study one or more of these subjects.

a) Draw a Venn diagram to illustrate the information.

b) How many pupils study all three subjects?

c) What is the probability that a pupil chosen at random studies all subjects mentioned?

10 Proportion/Variation

This section covers the following topics:

- Direct Proportion
- Inverse Proportion

LEARNING OBJECTIVES

By the end of this unit, you should be able to:

a) Write equations to solve direct proportion problems
b) Use equations to solve inverse proportion problems

KEYWORDS

- Direct proportion
- Inverse proportion
- Constant of proportionality

10.1 DIRECT PROPORTION

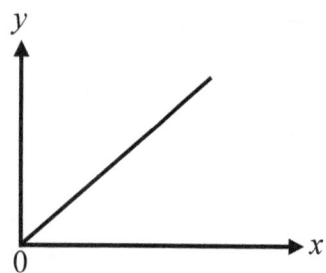

From the above graph, y is proportional to x. The above statement can be written as $y \propto x$.

If the slope of the line is **k**, then $y = \mathrm{k}x$. The quantity k is called the **constant of proportionality.**

Rearranging the equation gives, $k = \dfrac{y}{x}$

Therefore, the ratio of y to x is always a constant. From the above, y is said to vary **directly** as x.

It follows that x increases in equal amounts as y.

x also decreases in equal amounts as y. For example, if x is doubled, y is also doubled. A glaring example is a relationship between the circumference (C) of a circle and its diameter (d).

$C = \pi \times$ diameter

It means that the circumference of a circle is directly proportional to its diameter.

Example 1:
y is directly proportional to x.
When $y = 12$, $x = 3$. Calculate the value of a) y when $x = 7$ b) x when $y = 20$.

Solution:
From the question, $y \propto x$. To introduce the equality sign, multiply x by the constant of proportionality, k.

This gives $y = kx$ .... (1)
Using the values given for y and x,
$12 = k \times 3$ and $k = 4$.

Rewrite equation 1 with the value of k.

$$\boxed{y = 4x} \quad \ldots\ldots\ldots\ldots\ldots\ldots (2)$$

Using equation 2,
a) $y = 4x$
   $y = 4 \times 7 = \textbf{28}$

b) From $y = 4x$, $20 = 4 \times x$
   $x = \dfrac{20}{4} = \textbf{5}$

Example 2: The volume (V) of a solvent in a jar is proportional to the height (h) of the jar. When the height of the solvent in the jar is 12 cm, the volume is 45 cm³. Work out the volume of solvent when the height of solvent is 8 cm.

Solution: $V \propto h$ and $V = kh$

$45 = k \times 12$ and $k = \dfrac{45}{12} = 3.75$.
Rewrite the equation to give
$V = 3.75h$,
$V = 3.75 \times 8 = 30$

Volume of solvent = **30 cm³**.

Example 3:
B varies directly as x^2.
When B = 75, x = 5. Work out
a) B when x = 4   b) x when B = 300.

Solution: B $\propto$ x^2 and B = kx^2

Using the given numbers,
$75 = k \times 5^2$ and $k = \dfrac{75}{25} = 3$
Rewrite the equation to give
B = $3x^2$

a) B = $3 \times 4^2 = 3 \times 16 = $ **48**

b) From B = $3x^2$,
$$300 = 3 \times x^2$$
$$\frac{300}{3} = x^2$$

$$100 = x^2$$
$$x = \sqrt{100}$$
$$x = 10$$

EXERCISE 10A

1) W varies directly as T. If W = 30 and T = 6, calculate the value of a) W when T = 9   b) T when W = 17.5.

2) y is directly proportional to x. When y = 20 and x = 8.

a) Write an equation in terms of y and x.

b) Calculate the value of

i) y when x = 0.6   ii) x when y = 40.

3) If c is directly proportional to d, complete the table below.

c	15		27	
d	5	7		10.5

4) e varies directly as the square of f.
Given that e = 13.5 when f = 1.5,

a) form an equation in terms of e and f.
b) Calculate the value of i) e when f = 4 and ii) f when e = 2.

5) y is in direct proportion to the cube of x. Given that y = 320 when x = 4, calculate
a) y when x = 2.5   b) x when y = 53.24.

6) a is directly proportional to the square root of b. Complete the table.

a		10	16	
b	1	25		100

7) p varies in direct proportion to the cube root of q. When p = 6, q = 8.
a) Form an equation in terms of p and q.
b) Calculate (i) p when q = 125
ii) q when p = 9.

10.2 INVERSE PROPORTION

If one quantity increases at the same rate as the other decreases, then the two quantities are in **inverse proportion**. If one quantity doubles, the other one halves.

We could say that y is inversely proportional to x or that y varies as the inverse of x.

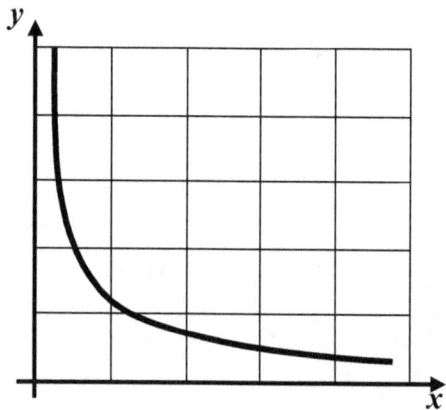

The above graph is a typical graph of y is inversely proportional to x.

We write it as:

$$\mathbf{y} \propto \frac{1}{x}$$

Removing the proportional sign and adding the constant of proportionality gives $y = k \times \dfrac{1}{x}$

$$y = \frac{k}{x}$$

Example 1:
y varies as the inverse of x.
When x = 5, y = 2.
a) Write the equation linking x and y.
b) Work out the value of
i) y when x = 4   ii) x when y = 8.

Solution: $y \propto \dfrac{1}{x}$

To remove the proportionality sign, we multiply by the constant of proportionality, k.

$$y = \frac{k}{x}$$

Using the values given and substituting,
$2 = \dfrac{k}{5}$ and k = 2 × 5 = 10.

a) The equation is $\mathbf{y = \dfrac{10}{x}}$

b) i) $y = \dfrac{10}{4} = \mathbf{2.5}$

ii) $8 = \dfrac{10}{x}$ and $x = \dfrac{10}{8} = \mathbf{1.25}$

Example 2: w is inversely proportional to t^2. When w = 5, t = 2.
a) Form an equation linking w and t.
b) Work out
i) w when t = 2.5   ii) t when w = 4

Solution: $w \propto \dfrac{1}{t^2}$ so $w = \dfrac{k}{t^2}$

$5 = \dfrac{k}{2^2}$ so k = 5 × 2^2 = 5 × 4 = 20

a) the equation is $\mathbf{w = \dfrac{20}{t^2}}$

b) i) $w = \dfrac{20}{t^2} = \dfrac{20}{2.5^2} = \dfrac{20}{6.25} = \mathbf{3.2}$

ii) $4 = \dfrac{20}{t^2}$ and $4t^2 = 20$

$t^2 = 20 \div 4 = 5$

$t = \mathbf{\sqrt{5}}$ or **2.24** to two decimal places.

EXERCISE 10B

1) y is inversely proportional to x.
When $y = 8$, $x = 3$.
a) Write an equation for y in terms of x.
b) Calculate the value of y when $x = 4$.
c) Calculate the value of x when $y = 10$.

2) w is inversely proportional to c.
When $w = 5.5$, $c = 2.5$.
a) Write an equation for w in terms of c.
b) Calculate the value of w when $c = 10$.
c) Calculate the value of c when $w = 8$.

3) Match each statement to the correct graph shown below.

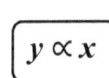 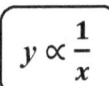 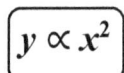

a)
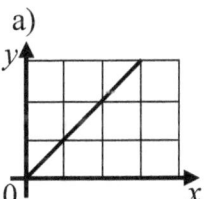

b)
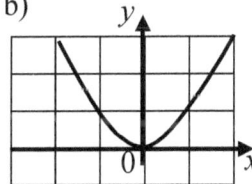

c)
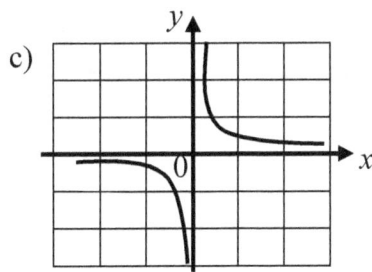

4) b is inversely proportional to c.
Complete the table.

b	1	6		15
c		2	4	

5) Below is a graph that shows two variables p and q that are inversely proportional to each other.
Find c, d and e.

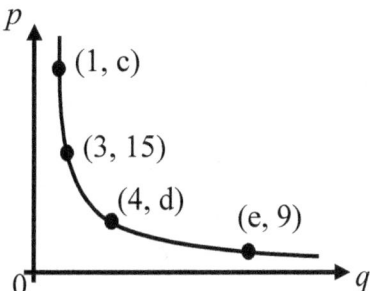

6) y is inversely proportional to $\sqrt{x}$.
When y = 50, x = 16.

a) Form an equation in term of y and x.

b) Calculate the value of y when x = 36.

c) Calculate the value of x when
y = 18.75.

7) When oil is poured into a cylinder, the depth, d of the oil is inversely proportional to the square of the radius, r of the cylinder.

When $r = 7$ cm, $d = 20$ cm.

a) Write a formula for d in terms of r.

b) Calculate the depth of the oil in the cylinder when the radius is 4 cm.

c) Calculate the radius of the cylinder when the depth is 115 cm to 1 d.p.

11 Other Non-Linear Graphs

This section covers the following topics:

- Cubic graphs
- Reciprocal graphs
- Exponential graphs
- Trigonometric graphs
- Trigonometric equations
- Transforming graphs

LEARNING OBJECTIVES

By the end of this section, you should be able to:

- Draw and interpret cubic graphs
- Draw and interpret reciprocal graphs
- Draw and interpret exponential graphs
- Draw trigonometric graphs of sin, cos and tan
- Solve trigonometric equations
- Transform graphs

KEYWORDS

- Cubic graphs
- Exponential graphs
- Reciprocal graphs
- Trigonometric equations
- Transforming graphs

11.1 CUBIC AND RECIPROCAL GRAPHS

A **cubic** function has the highest index number as 3. It will always include a power of 3. Examples of cubic numbers include: $x^3 + 4x + 8$, $3x^3 - 2x + 5$.

Example 1: a) Plot the graph of $y = x^3 + 2x + 1$, x values from -4 to 4.
b) Use your graph to find the value of y when x = 2.5.

Solution: Draw a table of values for the coordinates.
When $x = -4$, $y = (-4)^3 + 2(-4) + 1 = -71$

x	-4	-3	-2	-1	0	1	2	3	4
y	-71	-32	-11	-2	1	4	13	34	73

a)

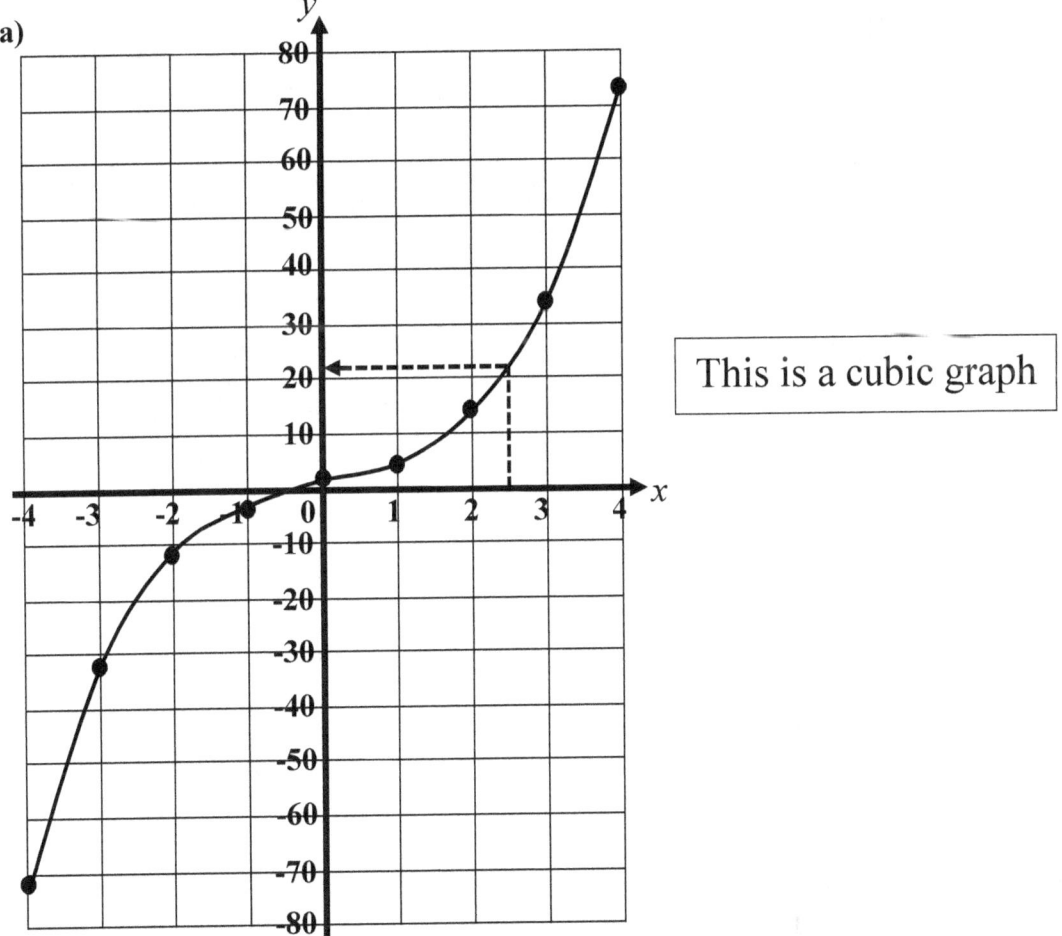

This is a cubic graph

b) From 2.5 on the x-axis, draw a vertical line to meet the graph. Then, read off the value on the y-axis. Therefore, when x = 2.5, y is ≈ **21.5**

RECIPROCAL GRAPHS

A **reciprocal** function is in the form $\frac{1}{a}$ or a^{-1}. $y = \frac{1}{x}$ and $y = \frac{5}{x}$ are examples of reciprocal functions. In a reciprocal function, the graph **never** touches the axes. It gets closer but not touching. A sketch of the reciprocal function of $y = \frac{1}{x}$ is shown below.

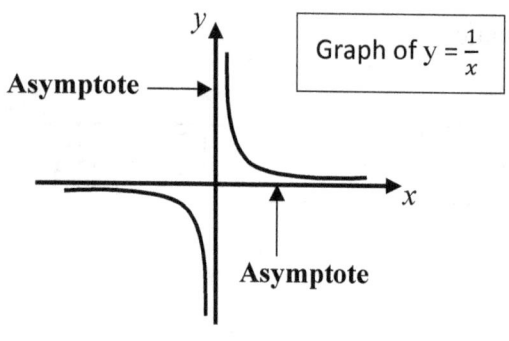

When a graph gets closer to a line but never touches it, that line is called an **asymptote**.

In a reciprocal graph, **two** asymptotes are visible.

For the graph of $y = \frac{1}{x}$, the equations of the asymptotes are $x = 0$ and $y = 0$.

Example 2: a) Draw the graph of $y = \frac{2}{x}$. Choose suitable values for the x-axis.

b) Write down the equations of the asymptotes.

Solution:

x	-5	-4	-3	-2	-1	-0.5	0.5	1	2	3	4
y	$-\frac{2}{5}$	$-\frac{1}{2}$	$-\frac{2}{3}$	-1	-2	-4	4	2	1	$\frac{2}{3}$	$\frac{1}{2}$

From $y = \frac{2}{x}$, when $x = 4$, $y = \frac{2}{4} = \frac{1}{2}$ or **0.5**

a)

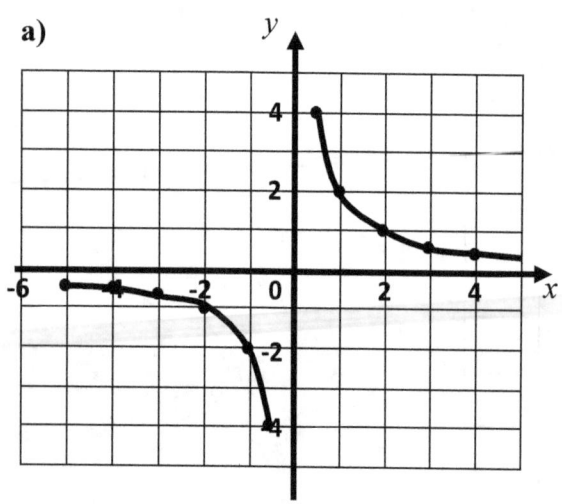

b) The asymptotes are: $x = 0$ and $y = 0$

EXERCISE 11A

1) Copy and complete the table of values for each equation.

a) $y = x^3 + 1$

x	-4	-3	-2	-1	0	1	2	3
y								

b) $y = x^3 + x - 3$

x	-4	-3	-2	-1	0	1	2	3
y								

c) $y = \dfrac{4}{x}$

x	-5	-4	-3	-2	-1	-0.5	0.5	1	2	3	4
y											

d) $y = -\dfrac{4}{x}$

x	-5	-4	-3	-2	-1	-0.5	0.5	1	2	3	4
y											

2) a) Draw the graphs of the functions in question **1a** and **1c** only.
 b) Show the asymptotes and write down their equations for **1c** only.

3) Copy and complete the table for $y = \dfrac{x}{x + 2}$.

x	-3	-2	-1	$-\frac{1}{2}$	$-\frac{1}{5}$	$\frac{1}{5}$	1	2	3
$x + 2$	-1			$1\frac{1}{2}$					5
$y = \dfrac{x}{x+2}$	3			-0.33					0.6

4) Copy and complete the table for the function $y = (x + 3)^3$.

X	-4	-3	-2	-1	0	1	2	3	4
Y		0				64		216	

11.2 EXPONENTIAL FUNCTIONS

An **exponential function** has the form $y = a^x$ where a is a positive constant and x can vary. Some examples of exponential functions are: $y = 5^x$ and $y = (0.25)^x$.
Two possible graphs (shapes) for exponential functions.

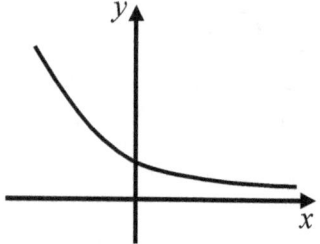

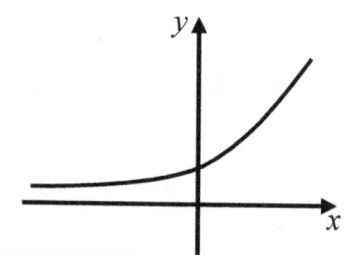

| x is increasing while y is decreasing (like a negative slope). The curve approaches the x-axis but never touches it. The x-axis is called the asymptote and the equation of the asymptote is $y = 0$. | x is increasing while y is increasing (like a positive slope). The curve approaches the x-axis but never touches it. The x-axis is called the asymptote and the equation of the asymptote is $y = 0$. |

Looking at the form of all exponential functions of the form $y = a^x$, it would suggest that the graph must pass through the coordinate point (0,1). This is explained as $y = a^0 = 1$.

Example 1: If $y = 4^x$ is an exponential function, a) draw its graph for x values from -3 to +4. b) Use your graph to estimate the value of x when y = 9.5.

Solution: Draw a table of values.

a)

x	-3	-2	-1	0	1	2
$y = 4^x$	0.02	0.06	0.25	1	4	16

b) From the graph, when y = 9.5, $x \approx 1.6$.

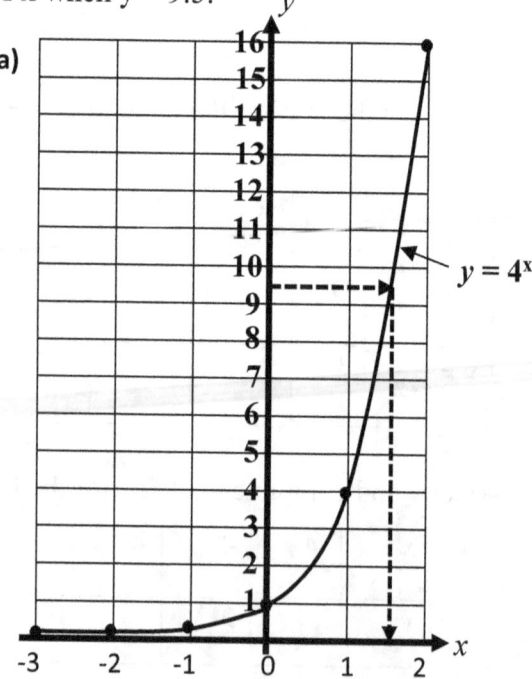

EXERCISE 11B

1) a) Copy and complete the table for $y = 3^x$.

x	-4	-3	-2	-1	0	1	2	3
$y = 3^x$	0.01						9	

b) Draw the graph of $y = 3^x$.

c) Use your graph to estimate the value of x when $y = 2.5$.

2) Draw a table of values for the graphs of a) $y = 5^x$ b) $y = 5^{-x}$. x values from -3 to 3.

c) At your leisure time, draw the graphs of $y = 5^x$ and $y = 5^{-x}$.

3) A radioactive substance decays by the exponential function

$$B = 60 \times 2^{-t}$$

where B is the amount of the radioactive material and t is the time in years.

a) Copy and complete the table below.

t	0	1	2	3	4	5	6	7	8
B		30						0.47	

b) Draw the exponential graph of the decay over the eight years.

c) Use the graph to estimate the amount of radioactive material remaining after $4\frac{1}{2}$ years.

4) a) Copy and complete the table for the function $y = (\frac{2}{5})^x$.

x		-2	-1	0	1	2	3	4
$y = (0.4)^x$			2.5		0.4			

b) At your leisure time, draw the graph of $y = (\frac{2}{5})^x$, for $-2 \leq x \leq 4$.

11.3 SINE, COSINE AND TANGENT GRAPHS

This section deals with recognising and drawing the graphs of $y = \sin \theta$, $y = \cos \theta$, and $y = \tan \theta$.

THE SINE GRAPH ($y = \sin \theta$)

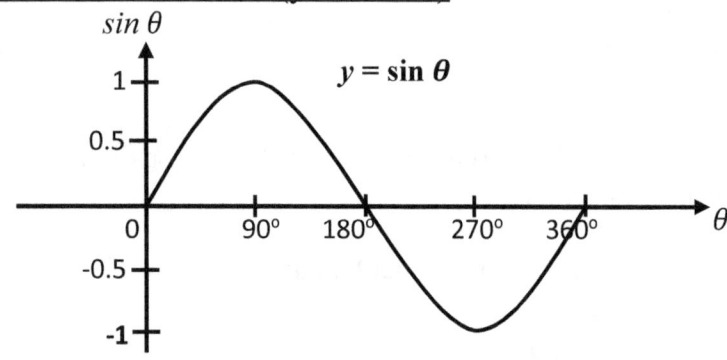

CHARACTERISTICS OF THE SINE GRAPH

1) The highest value of y is **+1**. It occurs when $\theta = 90°, 450°, -270°, \ldots$
2) The minimum value of y is **-1**. It occurs when $\theta = -90°, 270°, -450°, \ldots$
3) The graph repeats itself every 360°. It shows a **period** of 360°.

Below is the extended graph of $y = \sin \theta$ to show other values after each period.

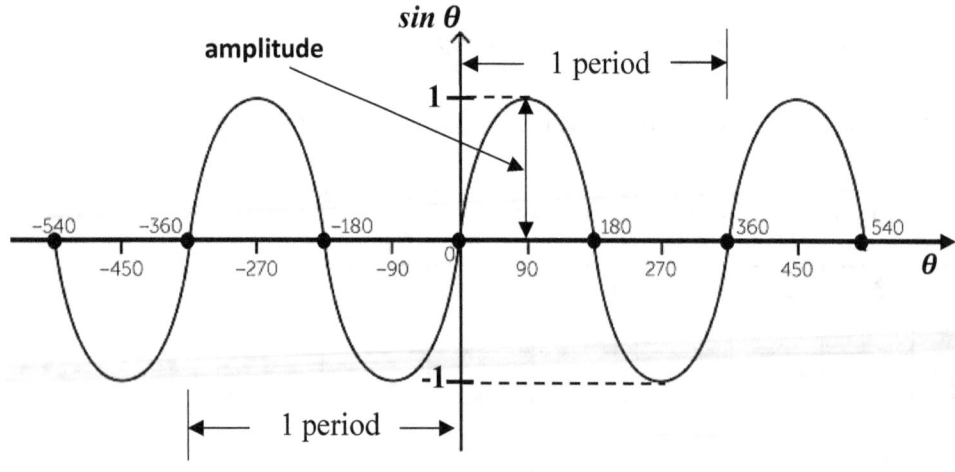

POINTS TO NOTE: $\sin(-\theta) = -\sin\theta$, $\sin(180° + \theta) = -\sin\theta$ and $\sin(180° - \theta) = \sin\theta$.

THE COSINE GRAPH (y = cos θ)

Below is a graph of $y = \cos\theta$. It has a maximum value of **+1** and a minimum value of **-1** just like the sine graph.

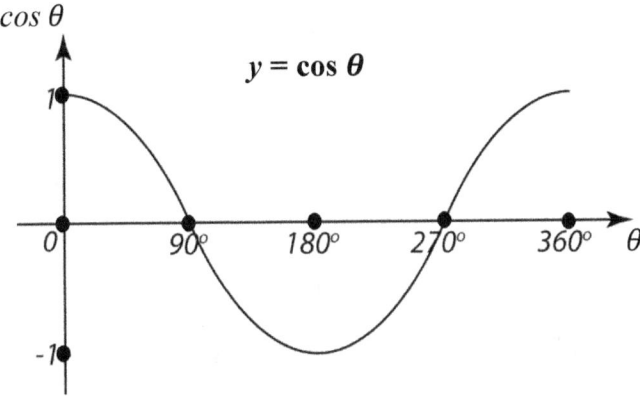

The cosine graph repeats itself every 360° and its **period** is 360°. The maximum value of +1 occurs when $\theta = 0°$, 360°, -360°, ... and the minimum value of -1 occurs when $\theta = 180°$, -180°, 540°, ...

THE TANGENT GRAPH (y = tan θ)

The graph will never touch the 90° line. Type in tan 90° in your calculator. What do you notice? This topic will be explained in detail in subsequent Whiz-kid, Mathematics Series (A levels).

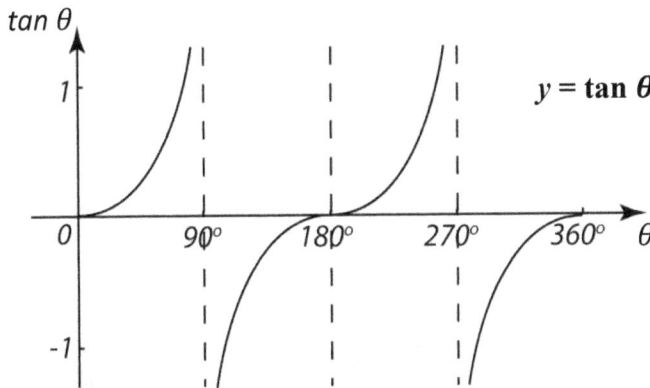

The tan graph repeats itself every 180o and the period is 180o. It is also infinite at $x(\theta) = \pm 90°, \pm 270°, ...$

EXERCISE 11C

1) a) Copy and complete the table and give your values to two decimal places.

θ	0°	30°	45°	60°	90°	135°	180°
sin θ							
cos θ							
tan θ							

b) Draw the graph of $y = \sin \theta$ for x values in the interval $0° \leq \theta \leq 360°$.
c) Draw the graph of $y = \cos \theta$ for x values in the interval $0° \leq \theta \leq 360°$

2) From the graph of $y = \sin \theta$ in question **1b**, estimate the angle when
i) $\sin \theta = 0.5$ ii) $\sin \theta = 0.87$

3) From the graph of $y = \cos \theta$ in question **1c**, estimate the angle when
i) $\cos \theta = 0.5$ ii) $\cos \theta = 0.86602$.

4) a) Copy and complete the table.

x	$2x$	$2 \sin 2x$
-180°		
-135°	-270°	2
-90°		
-45°		
0°		
45°		
90°	180°	
135°		
180°		0

b) Draw the graph of $y = 2 \sin 2x$.
c) Work out the minimum and maximum values of $2\sin 2x$ and the values of x at which they occur.

11.4 TRIGONOMETRIC EQUATIONS

To any trigonometric equation, there is uncountable (infinite) number of solutions due to the symmetry and properties of the curves.

Using a calculator will give only one value and students should respond accordingly when prompted.

Example 1:
a) Solve the equation $10 \sin x = 5$.
b) Also, state other values of x in the interval $0° \leq \theta \leq 360°$.

Solution:

a) $10 \sin x = 5$

$\text{Sin } x = \dfrac{5}{10} = \dfrac{1}{2}$ or 0.5.

$x = \sin^{-1} 0.5 = 30°$

b) Using the symmetry of the sine curve $y = \sin x$,

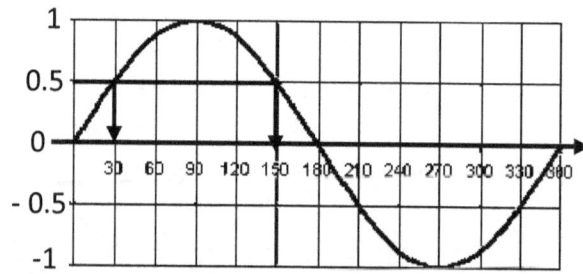

From 0.5, draw a horizontal line to cut the graph at two points. Read off the values to give **30°** and **150°**.

Check with a calculator. Sin 30° and 150° is $= 0.5$.

EXERCISE 11D

1) Solve the equation $4 \sin x = 2$ and give all the values of x in the range $-360° \le x \le 360°$.

2) a) Sketch the graph of $y = 4 \cos x$, x values from $0°$ to $360°$.

b) Solve the equation $4 \cos x = 1$ and give all the values of x between $0°$ and $360°$.

3) For the interval, $0° \le \theta \le 360°$, find all the solutions of the equations

a) $\sin \theta = 0.6$ c) $\sin \theta = -0.2$

b) $\cos \theta = 0.6$ d) $\cos 0.965$

4) a) Solve the trigonometric equation $8 \sin 2x = 3$, giving your answer to the nearest degree.

b) Find two solutions to the equation $8 \sin 2x = 3$ in the interval $0° \le \theta \le 180°$.

5) a) Use the graph to find i) $\cos 135°$ ii) $\cos 27°$.   b) Solve the equation $\cos x = 0.4$.

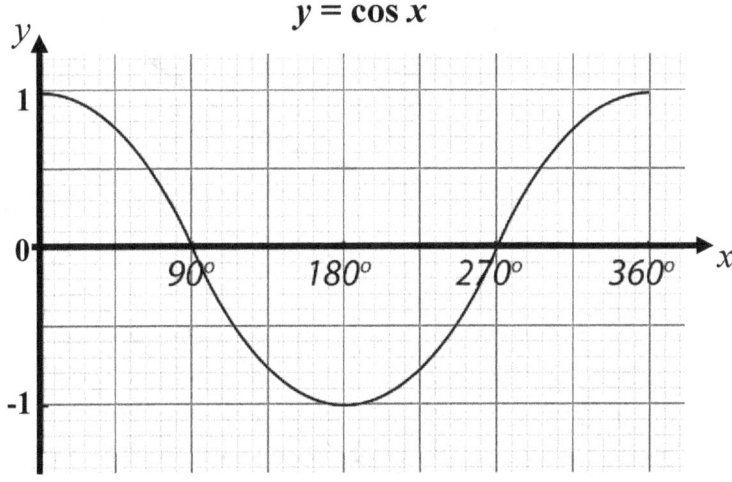

$y = \cos x$

11.5 TRANSFORMING GRAPHS

You should have encountered basic transformations (reflections, rotations, translations and enlargements) in earlier years as a mathematics student.

This section deals with translations and its applications when manoeuvring from one function to another. It will involve some stretches along the y and x-axes depending on the functions to be translated/transformed. For ease of understanding, the graph of $y = x^2$ will be used to demonstrate different transformations. However, the rules apply to **any shape** or function to be transformed.

RULES FOR TRANSFORMATIONS

1) $\mathbf{f(x) + c}$ moves the function **c** units upwards. See graph ($y = x^2 \longrightarrow y = x^2 + 3$)

The graph of $y = x^2$ has moved vertically upwards 3 units by the vector $\binom{0}{3}$.

2) $\mathbf{f(x) - c}$ moves the function **c** units downwards. See graph ($y = x^2 \longrightarrow y = x^2 - 8$)

The graph of $y = x^2$ has moved down 8 units by the vector $\binom{0}{-8}$.

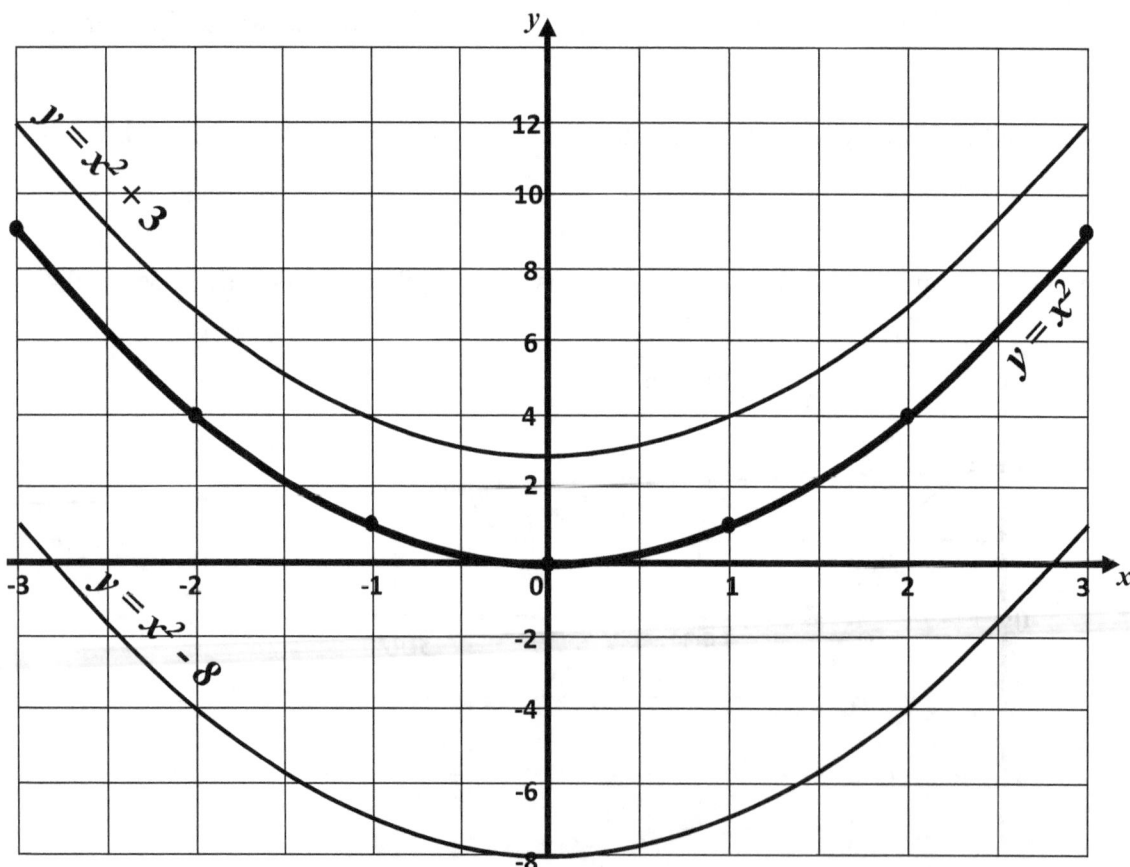

3) **f (x + c)** moves the function **c** units to the left. From the diagram below, the graph of $y = x^2$ is translated 1 unit to the **left** by the vector $\begin{pmatrix} -1 \\ 0 \end{pmatrix}$ to become $y = (x + 1)^2$.

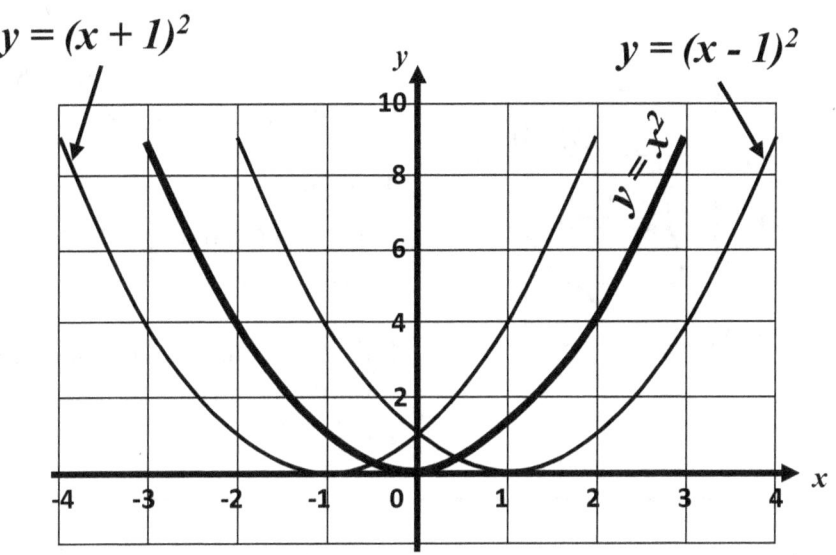

$y = (x + 1)^2$ $y = (x - 1)^2$

4) **f (x – c)** moves the function **c** units to the right. From the diagram above, the graph of $y = x^2$ is translated 1 unit to the **right** by the vector $\begin{pmatrix} 1 \\ 0 \end{pmatrix}$ to become $y = (x – 1)^2$.

5) If a function is multiplied by a constant, we talk of **stretches** instead of translations.

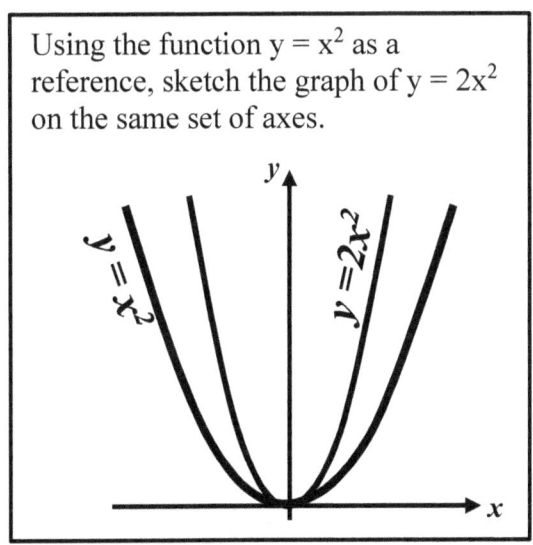

Using the function $y = x^2$ as a reference, sketch the graph of $y = 2x^2$ on the same set of axes.

The curve of $y = x^2$ has been stretched out in the direction of the y-axis by a scale factor of 2 to give the graph of $y = 2x^2$.

The y-coordinates of $y = 2x^2$ are 2 times bigger when compared with the y- coordinates of $y = x^2$.

Note to students: If in doubt about the shape of the transformed function, draw the graph by choosing sensible x and y values.

Also, the function $y = x^2$ can be reduced horizontally using a fractional scale factor. Again, using $y = x^2$ as a reference, sketch the graph of $y = (4x)^2$.

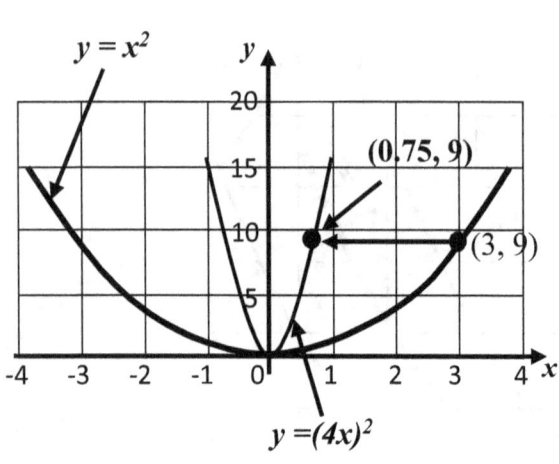

$y = x^2$

$y = (4x)^2$

(0.75, 9)

(3, 9)

The diagram is the sketch of $y = (4x)^2$. $(4x)^2 = 4x \times 4x = 16x^2$. The y-coordinates of $y = x^2$ has been multiplied by 16 to give the corresponding y-coordinates of $y = (4x)^2$.

The implication is that the x-coordinates of $y = x^2$ is a quarter $(\frac{1}{4})$ compared to that of $y = (4x)^2$.

It therefore moves the graph of $y = x^2$ inwards (towards the y-axis) to produce the graph of $y = (4x)^2$.

EXERCISE 11E

1a) Draw the graph of $y = x^2$, x values From -3 to 3.

b) On the same graph, draw the graph of
i) $y = x^2 + 1$ ii) $y = x^2 - 3$ iii) $y = (x - 2)^2$

c) Describe the transformations that moves the graph of $y = x^2$ to the graphs in question **1b**.

2a) Sketch the graph of $y = x^2$.
b) Sketch the graph of $y = (x + 0.5)^2$.
c) Describe the transformation that takes the graph of **a** to the graph of **b**.

3) Sketch the graphs of the functions
a) $y = x^2$ and b) $y = 3x^2$ on the same graph.
c) Explain in full how the graph of $y = 3x^2$ is obtained from that of $y = x^2$.

4) Explain the transformation that takes the graph of $y = x^2$ to a) $y = 6x^2$ b) $y = (2x)^2$.

5) When translated by the vectors
a) $\begin{pmatrix} 0 \\ 2 \end{pmatrix}$ b) $\begin{pmatrix} 0 \\ -5 \end{pmatrix}$ c) $\begin{pmatrix} 1 \\ 0 \end{pmatrix}$,

the graph of $y = x^2$ moves to a different position. Sketch the graphs to illustrate the movements.

6) Explain fully the transformations that will take the function $y = f(x)$ to
a) $y = f(x) - 3$ b) $y = 3f(x) + 2$

7) The graph of $y = \sin x$ is shown below.

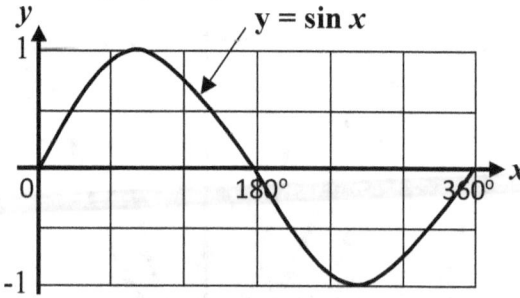

$y = \sin x$

180°

360°

a) Sketch the graphs of
i) $y = \sin x + 1$ ii) $y = \sin x - 2$
iii) $2 \sin x + 2$
b) What is the period of $y = \sin x + 1$?

90

ANSWERS

EXERCISE 1

1a) $x^3 + 7x^2 + 14x + 8$
b) $x^3 + 6x^2 + 11x + 6$
c) $x^3 - 9x^2 + 24x - 20$
d) $bcd + 3bd - c^2d - 3cd - 2bc - 6b + 2c^2 + 6c$
e) $w^3 - 10w^2 + 26w - 35$

2a) $30x^3 + 116x^2 + 82x + 12$
b) $125x^3 - 300x^2 + 240x - 64$

3a) $24w^3 + 38w^2 - 112w - 30$
b) $70w^2 - 8w - 38$
c) 42 cm by 8 cm by 10 cm

4) $g = 6$ and $c = 1$

EXERCISE 2A

1a) $x = -3$ or $x = -\frac{1}{5}$

b) $x = -4$ or $x = -3.5$
c) $x = -1.5$ or $x = -\frac{4}{5}$

d) $x = -10$ or $x = -\frac{1}{4}$

e) $x = -5$ or $x = -\frac{1}{6}$

2a) $x = -\frac{3}{4}$ or $x = 2$

b) $x = 1.5$ or $x = \frac{5}{3}$

c) $x = 2$ or $x = \frac{5}{4}$

d) $x = \frac{3}{7}$ or $x = -\frac{5}{6}$

e) $x = -1$ or $x = \frac{1}{3}$
3a) $a = 2$ and $- 2$
b) $a = 3$ and -3
c) $x = 10$ and $- 10$
d) $x = 12$ and $- 12$
e) $x = 15$ and $- 15$
f) $r = 1$ or -1
g) $k = 9$ or $- 9$
h) $w = 13$ or $- 13$
i) $w = 18$ or $- 18$
j) $m = 27$ or $- 27$
4a) $w = 1$ and $w = - 5$

b) $w = 12$ and $w = - 4$
c) $w = 6$ and $w = - 10$
d) $w = 8$
e) $w = 6$ and $w = 7$

EXERCISE 2B

1) $x = \dfrac{-9 \pm \sqrt{73}}{2}$

2) $x = \dfrac{-4 \pm \sqrt{12}}{2}$

3) $x = \dfrac{-7 \pm \sqrt{73}}{2}$

4) $x = \dfrac{-5 \pm \sqrt{37}}{2}$

5) $x = \dfrac{3 \pm \sqrt{17}}{2}$

6) $x = \dfrac{-7 \pm 5}{4}$

7) $x = \dfrac{-8 \pm \sqrt{88}}{6}$

8) $x = \dfrac{5 \pm \sqrt{97}}{12}$

9) $x = \dfrac{-5 \pm \sqrt{73}}{8}$

10) $x = \dfrac{1 \pm \sqrt{29}}{2}$

EXERCISE 2C

1) $x = 0.772$ and $x = - 7.77$

2a) $x = 3.19$ or $x = -2.19$
b) $x = 2.11$ or $x = - 2.61$
c) $x = - 1.70$ or $x = - 5.30$
d) $y = 5.54$ or $y = - 0.541$
e) $x = 5.34$ or $x = - 0.0936$
f) $x = 12.3$ or $x = 5.68$

3) The two numbers are 0.65 and 7.65

4) 17 and 18

5) 5 m

EXERCISE 2D

1) $b^2 - 4ac < 0$, therefore, no solution.

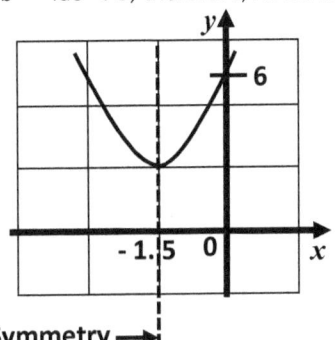

2) $b^2 - 4ac > 0$, therefore, two solutions.

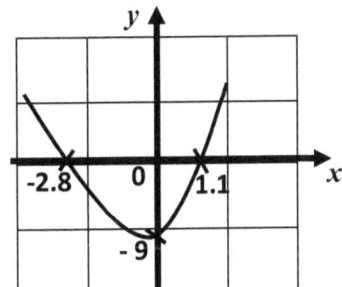

3) $b^2 - 4ac > 0$, therefore, two solutions.

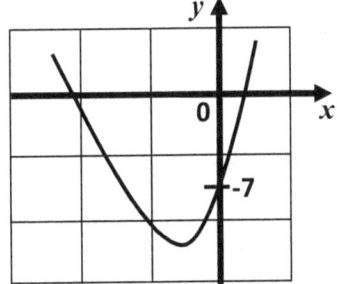

4) $b^2 - 4ac = 0$, therefore, one solution.

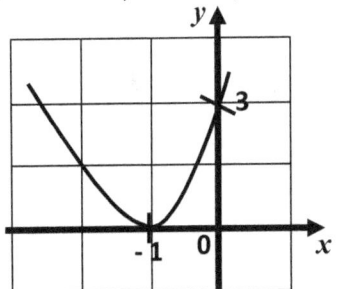

5) $b^2 - 4ac = 0$, therefore, one solution.

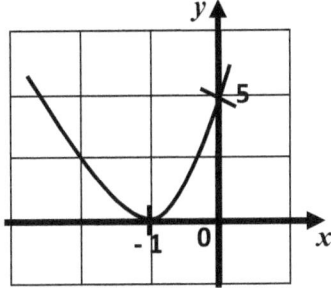

6) $b^2 - 4ac < 0$, therefore, no solution.

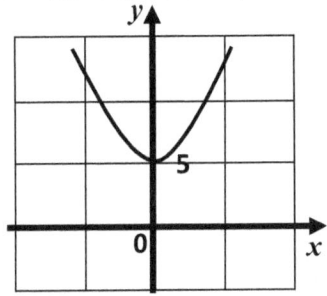

7) $b^2 - 4ac > 0$, therefore, two solutions.

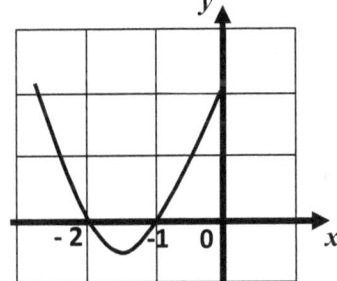

8) $b^2 - 4ac > 0$, therefore, two solutions.

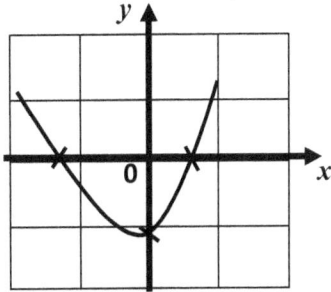

9a) P = - 3.56 and Q = 0.56

9b) x = - 1.5

93

EXERCISE 2E

1) $(x + 4)^2 - 14$
2) $(x + 5)^2 - 18$
3) $(x + 3)^2 - 18$
4) $(x - 3)^2 - 10$
5) $(x - 8)^2 - 59$
6) $(x + 1)^2 - 6$
7) $(x + 2.5)^2 - 1.25$
8) $(x - 3.5)^2 - 21.25$
9) $(x + 10)^2 - 93$
b = 10 and c = -93
10) $(x - 3.5)^2 - 7.25$
b = -3.5 and c = -7.25
11) $(x - 1.5)^2 - 8.25$
b = -1.5 and c = - 8.25
12) $(x + 7)^2 - 41$
b = 7 and c = -41
13) $(x + 3)^2 - 7$
b = 3 and c = -7
14) $(x - 2)^2 - 8$
b = -2 and c = -8
15) $(x + 1)^2 + 4$
b = 1 and c = 4
16) $(x + 3)^2 + 4$
b = 3 and c = 4
17) p = 4 and q = 24

EXERCISE 2F

1a) $m = \pm \sqrt{19} - 4$
b) $x = \pm\sqrt{21} - 4$
c) $x = \pm\sqrt{5} + 2$
d) $x = \pm\sqrt{19} - 5$
e) $x = \pm\sqrt{11} + 3$
f) $x = \pm\sqrt{27} - 6$

2a) c = 1 or c = -3
b) x = 0.58 or x = -8.58
c) x = 4.24 or x = -0.24
d) x = -0.64 or x = -9.36
e) x = 6.32 or x = -0.32
f) y = -0.80 or y = -11.20

EXERCISE 2G

1) $3(x + 1)^2 + \frac{2}{3}$

2) $5(x + 1.5)^2 - 1.25$
3) $4(x - 0.75)^2 - 1.25$
4) $7(x - 1)^2 - 1$
5) $0.5(n + 5)^2 - 6.5$
6) $0.75(k - 2)^2 + 3$

EXERCISE 2H

1) x = -1 or -2
2) x = 1.31 or 0.19
3) x = 1.38 or 0.62
4) x = -1.39 or -8.61

EXERCISE 2I

1) $2(x + 1.5)^2 + 2.5$
2) $3(x - 1.5)^2 - 5.75$
3) $4(b + 0.5)^2 + 2$
4) $7(c - 1.5)^2 - 7.75$
5) $5(x - 1)^2 - 10$
6) $9(x + 0.5)^2 - 7.25$

7)
1) a = 2, p = 1.5, q = 2.5
2) a = 3, p = -1.5, q = -5.75
3) a = 4, p = 0.5, q = 2
4) a = 7, p = -1.5, q = -7.75
5) a = 5, p = -1, q = -10
6) a = 9, p = 0.5, q = -7.25

8i)

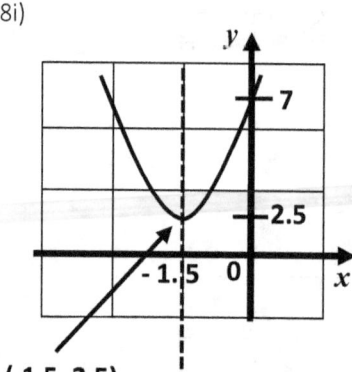

(-1.5, 2.5)

8ii)

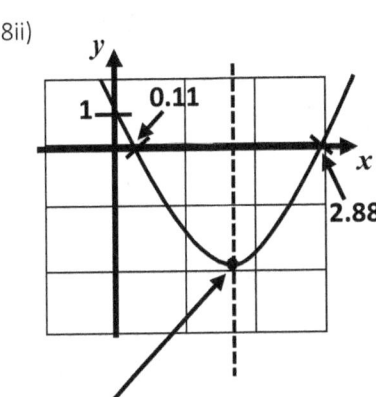

(1.5, -5.75)

9i)
1) (-1.5, 2.5)
2) $(1.5, -\frac{23}{4})$
3) (-0.5, 2)
4) 1.5, -7.75)
5) (1, -10)
6) (-0.5, 7.25)

9ii)
1) x = -1.5
2) x = 1.5
3) x = -0.5
4) x = 1.5
5) x = 1
6) x = -0.5

10a) $y^2 - 12y + 20 = 0$
b) y = 2, y = 10
c) 10 cm, 8 cm and 6 cm

11a) No solution
b) x = 0.11 or 2.89
c) No solution
d) x = 0.45 or 2.55
e) x = -0.41 or 2.41
f) x = 0.4 or -1.4

12a) c = -1 and d = 6
b) x = 2.414 and x = -0.414
c) (1, 6)
d) x = 1

EXERCISE 3

1) 9, 21, 45, 93, 189

2a) $x^3 + 7x - 3 = 0$
$x(x^2 + 7) - 3 = 0$
$x(x^2 + 7) = 3$
$x = \dfrac{3}{x^2 + 7}$

2b) 0.42

2c) $x^3 + 7x - 3 = 0$
$x^3 + 7x = 3$
$x^3 = 3 - 7x$
$x = \sqrt[3]{3 - 7x}$

3a) $y(y - 3)$
b) $y^2 - 3y - 64 = 0$
c) $y_{n+1} = \sqrt{3y_n + 64}$
d) $y = 9.64$

4a) $x^3 - 8x + 7 = 0$
$x^3 + 7 = 8x$
$x^3 = 8x - 7$
$x = \sqrt[3]{8x - 7}$

4b) $x_{n+1} = \sqrt[3]{8x_n - 7}$
4c) 2.19
5) $x = 3.19$
6) Does not exist

EXERCISE 4A

1) x < 6 and x > −6

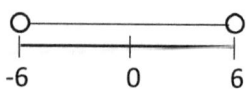

2) x < 2 and x > −2

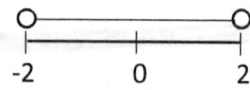

3) x < 10 and x > −10

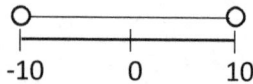

4) x > 6 and x < −6

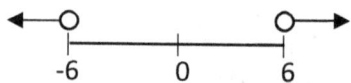

5) x > 2 and x < −2

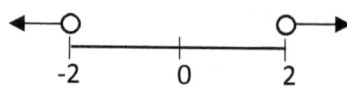

6) x > 10 and x < −10

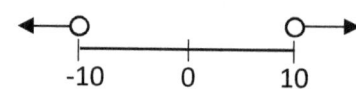

7) x ≤ 8 and x ≥ −8

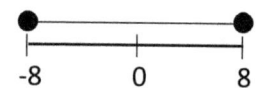

8) x ≤ 7 and x ≥ −7

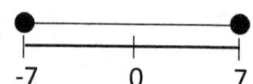

9) x ≤ 10 and x ≥ −10

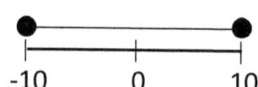

10) x ≥ 8 and x≤ −8

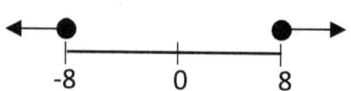

11) x ≥ 1 and x≤ −1

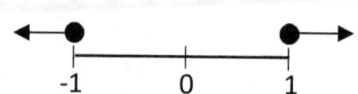

12) x ≥ 10 and x≤ −10

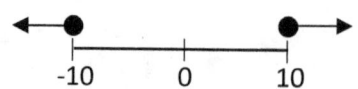

EXERCISE 4B

1) x < 4 and x > −4

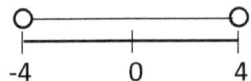

2) x < 6 and x > −6

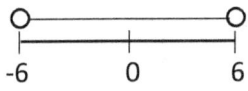

3) x > 8 and x< −8

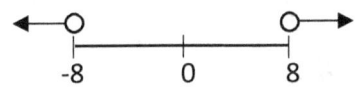

4) x ≤ 7 and x ≥ −7

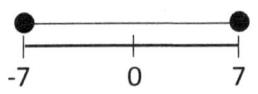

5) x < 9 and x > −9

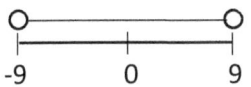

6) x ≤ 4 and x ≥ −4

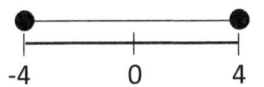

7) x > 4 and x< −4

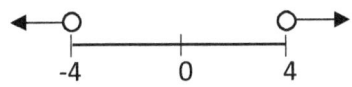

8) x ≤ 10 and x ≥ −10

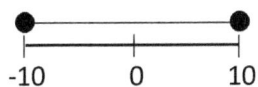

9) x < 5 and x > −7

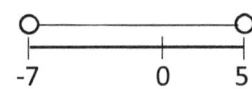

10) x ≥ 1 and x≤ −6

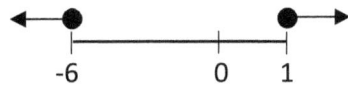

11) x ≤ 10 and x ≥ −3

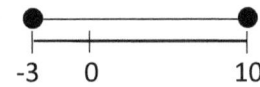

12) x > 4 and x< −5

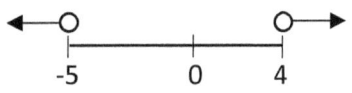

13) x < 4 and x > −8

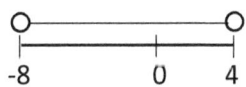

97

EXERCISE 5A

1a) 24   b) -11   c) 4   d) 17
2a) 4   b) -1   c) -1.8   d) -2.95
3a) 275   b) -200   c) 90
d) 18.75
4a) 27   b) 195   c) 3   d) 1
e) ± 4 5a) 123   b) -1697
c) 123   d) -3
6) 11, 12, 13, 14

EXERCISE 5B

1a) $x + 7$
b) $\frac{x-9}{2}$
c) $\sqrt{x+1}$
d) $3x - 2$
e) $\frac{3-5x}{x}$
f) $\sqrt[3]{x+10}$

2a) 15   b) -0.5   c) 3   d) 22
e) -4.625
f) $\sqrt[3]{18}$ = 2.62 to 2 d.p.

3) $\frac{9x+7}{5-2x}$
4a) $\frac{4}{x-3}$

4b) $\frac{17x-3}{3} = \frac{17x}{3} - 1$
4c) $\frac{13}{x}$

4d) $\frac{q+px}{rx-p}$

5a) $\frac{5x+1}{4-2x}$

5b) $-\frac{13}{3}$

EXERCISE 5C

1a) $5x^2 - 16$
b) $25x^2 - 10x - 2$
c) $44 - 5x^2$
d) $25x - 6$

2a) $\frac{4}{3}$ b) 6   c) $\frac{2}{3}$ d) $-\frac{2}{9}$
3a) $49x - 168$ b) $27 - 7x$
c) $21 - 7x$ d) 18.6   e) 42

4) $2x^2 - 3x - 2$

EXERCISE 6

1a) Student's own drawing
b) Student's own drawing of a circle with radius of 3.
c) diameter = 6
d) x – axes: (-3, 0) and (3, 0)
y – axes: (0, -3) and (0, 3)

2a)

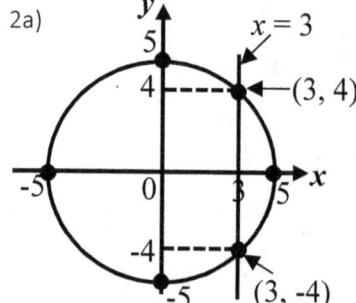

2b) See diagram above.
2c) $x^2 + y^2 = 25$
2d) x – axes: (5, 0) and (-5,0)
y – axes: (0, -5) and (0, 5)
2e) (3, 4) and (3, -4)

3a) Student's own drawing of a circle with radius of 8 units.
3b) Student's own drawing
3c) x = - 3.12 and x = 1.92
3d) 3
3e) Student's own drawing
3f) x = - 4.26 and x = 0.66

4a) $\frac{1}{2}$ 4b) $y = \frac{1}{2}x + 10$

5) y = 11 and x = 11

6a) $x^2 + y^2 = 9$
6b) $x^2 + y^2 = 169$
6c) $x^2 + y^2 = 5$
6d) $x^2 + y^2 = 27$
6e) $x^2 + y^2 = 15.21$

7a) $y = -\frac{7}{24}x + \frac{625}{24}$

7b) $y = \frac{7}{24}x + \frac{625}{24}$

8) $x^2 + y^2 = 40.44$

9a) $2\sqrt{17}$

9b) $y = -\frac{1}{4}x - 4.25$

9c) $y = \frac{3}{\sqrt{8}}x + \frac{17}{\sqrt{8}}$

10a) (0, 61) and (0, −61)
10b) (11, 60) and (11, −60)
10c) (−11, 60) and (11, 60)

EXERCISE 7A

1a) 12.5 m/s b) 25 m/s
c) 875 m d) 125 m
e) 17.5 m/s
2) 145 km/h
3a) 120 km
b) i) 25 km/h² ii) 6.7 km/h²
4a) 37.5 m/s
4b) P: Positive gradient with increasing velocity. Speed (velocity) increased from 0 m/s to 30 m/s for 10 seconds.
Q: This section is horizontal signifying that speed is constant at 30 m/s for 10 seconds (20 − 10 = 10).
R: Positive gradient with increasing velocity (30 m/s to 45 m/s) which is 15 m/s for 10 seconds (30 − 20= 10).
S: Decelerates for 20 seconds (50 − 30 = 20).
4c) Bike was decelerating for 20 seconds.
4d) $\frac{45}{20} = 2.25$ m/s²

EXERCISE 7B

1) 175 km and under-estimation. (Teacher's guidance recommended)
2a) 40 m/s b) 80 m/s
c) 330 m/s
3a) 190 km
b) under-estimation.
4a) 105 km
b) over-estimation
5) 192.5 m

EXERCISE 7C

1a) 10 hrs. This is the highest point on the curve before deceleration.
1b) i) 12.5 km/h ii) 5 km/h
1c) 12.5 km/h
2a) 2.7 m/s² 2b) 1.75 m/s²
2c) t = 20. This is the highest point on the curve before deceleration.
3a) 6.7 km/h² 3b) 365 km
4a) 15 km/h 4b) 0 km/h
4c) 13.333...km/h
5a)

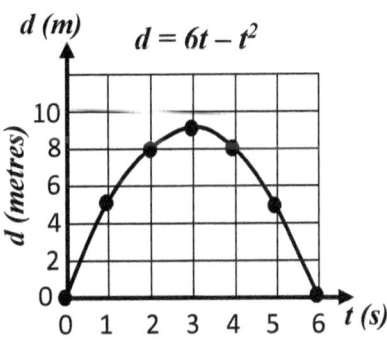

5b) The gradient is zero at the maximum point (3,9). Therefore, the time when gradient is zero is 3 seconds.

5c) Maximum speed is when t = 0 or at t = 6. This occurs when the curve is steepest. By drawing a tangent at t = 0, the gradient is 6. Therefore, the maximum speed is 6 m/s.

5d) At point (4, 8), the gradient is negative, as the line is going down. The gradient will be -2 and it can be said that the speed is moving down with a speed of 2 m/s.

6a) 0.5 m/s²

6b) Zero

6c) ≈ 1575 m (teacher's guidance recommended).

7a)

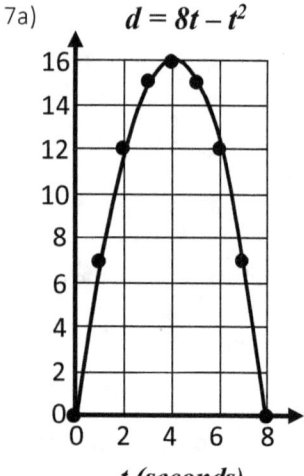

$$d = 8t - t^2$$

t (seconds)

7b) Maximum speed is when t = 0 or at t = 8.

$\frac{dy}{dx} = 8 - 2t$

When t = 0, speed (gradient) = 8 m/s.

7c) i) $\frac{dy}{dx} = 8 - 2t$

When t = 2, the gradient/speed = 4 m/s.

ii) $\frac{dy}{dx} = 8 - 2t$

when t = 4, speed = 0 m/s.

EXERCISE 8

1a) $\frac{1+\sqrt{3}}{2}$ b) $\frac{2}{\sqrt{3}}$ c) $\frac{3\sqrt{3}}{2}$

d) $1\frac{1}{\sqrt{2}}$ e) $9\frac{\sqrt{3}}{2}$ f) 3

g) $\frac{\sqrt{3}}{2} + \frac{1}{\sqrt{3}}$

$= \frac{\sqrt{3}(\sqrt{3})}{2 \times \sqrt{3}} + \frac{1 \times 2}{2\sqrt{3}} = \frac{3}{2\sqrt{3}} + \frac{2}{2\sqrt{3}}$

$= \frac{5}{2\sqrt{3}}$

Rationalise the denominator

$= \frac{5}{2\sqrt{3}} \times \frac{\sqrt{3}}{\sqrt{3}} = \frac{5 \times \sqrt{3}}{2\sqrt{3} \times \sqrt{3}}$

$= \frac{5\sqrt{3}}{6}$

2a) $\sqrt{3}$ b) $\frac{19\sqrt{2}}{2}$ c) 6.4

EXERCISE 9A

1a) P = {7,14,21,28,35,42}

b)

Q= {1,2,3,4,6,8,12,16,24,48}

c) R = {7,11,13,17,19,23}

d) S = {27,64}

2a) 35 pupils   b) 22 pupils

c) 14 pupils   d) 13 pupils

e) 5 pupils   f) $\frac{9}{35}$ g) $\frac{22}{35}$

h) $\frac{27}{35}$ i) $\frac{8}{35}$ j) $\frac{22}{35}$ k) $\frac{26}{35}$

3a)

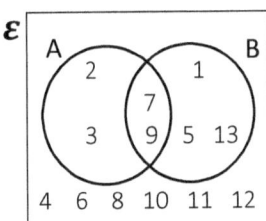

3b) i) $\frac{5}{13}$ ii) $\frac{9}{13}$ iii) $\frac{2}{13}$ iv) $\frac{6}{13}$

4a)

B = {2,4,6,8,10,12,14,16,18}

C = {4,8,12,16}

D = {2,3,5,7,11,13,17,19}

5a) $A' \cap B$

b) $A \cap B'$

c) $A \cap B$

d) $(A \cup B)'$

e) $A \cup B$

f) A'

EXERCISE 9B

1a) c = Cat and D = dog

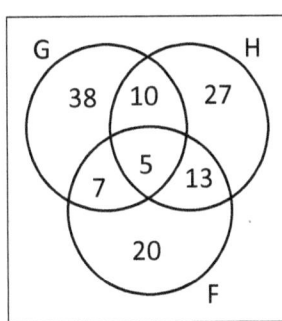

b) i) $\frac{21}{27} = \frac{7}{9}$ ii) $\frac{9}{27} = \frac{1}{3}$

c)) $\frac{8}{20} = \frac{2}{5}$

2a) 0.65   b) 0.88

3a)

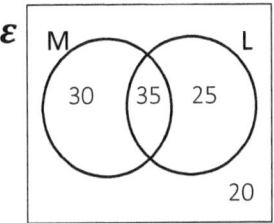

Where M = media dept and
L = left handed workers

b) 35 workers

c) $\frac{25}{110} = \frac{5}{22}$

4a) w = 15   b) 36   c) 40

5a) $\frac{5}{15}$ b) = $\frac{8}{15}$ c) $\frac{10}{15}$

d) $\frac{7}{15}$ e) $\frac{12}{15}$ f) $\frac{4}{15}$ g) $\frac{7}{15}$

6a) G = Geography,
H = History and F = French.

6b) 5 students   c) $\frac{5}{120}$

EXERCISE 10A

1a) 45   b) 3.5
2a) y = 2.5 x
2b) i) 1.5   ii) 16
3)

c	15	21	27	31.5
d	5	7	9	10.5

4a) $e = 6f^2$
b) i) 96   ii) 0.58 to 2 d.p.
5a) 78.125   b) 2.2

6)

a	2	10	16	20
b	1	25	64	100

7a) $p = 3\sqrt[3]{q}$
b) i) 15   ii) 27

EXERCISE 10B

1a) $y = \dfrac{24}{x}$ b) 6   c) 2.4
2a) $w = \dfrac{13.75}{c}$
2b) 1.375   c) 1.72
3) graph (a) $y \propto x$
graph (b) $y \propto x^2$
graph (c) $y \propto \dfrac{1}{x}$

4)

b	1	6	3	15
c	12	2	4	0.8

5) c = 45   d = 11.25   e = 5
6a) $y = \dfrac{200}{\sqrt{x}}$
6b) 33.333...   6c) 113.8
7a) $d = \dfrac{980}{r^2}$
7b) 61.25 cm   c) 2.9 cm

EXERCISE 11A

1a) $y = x^3 + 1$

x	-4	-3	-2	-1	0	1	2	3
y	-63	-26	-7	0	1	2	9	28

1b) $y = x^3 + x - 3$

x	-4	-3	-2	-1	0	1	2	3
y	-71	-33	-13	-5	-3	-1	7	27

1c) $y = \dfrac{4}{x}$

x	-5	-4	-3	-2	-1	$-\frac{1}{2}$	0.5	1	2	3	4
y	-0.8	-1	-1.3	-2	-4	-8	8	4	2	1.3	1

1d) $y = -\dfrac{4}{x}$

x	-5	-4	-3	-2	-1	$-\frac{1}{2}$	0.5	1	2	3	4
y	0.8	1	1.3	2	4	8	-8	-4	-2	-1.3	-1

2a) 1a

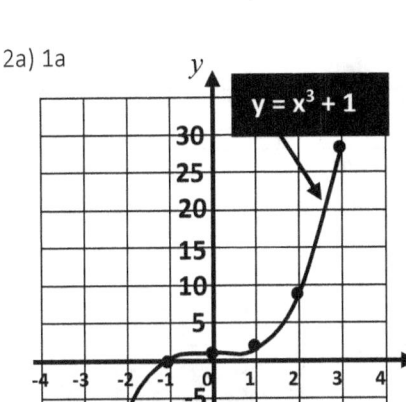

3) $y = \dfrac{x}{x+2}$

x	-3	-2	-1	-0.5	-0.2	0.2	1	2	3
x + 2	-1	0	1	1.5	1.8	2.2	3	4	5
y	3	-	-1	-0.33	-0.11	0.09	0.33	0.5	0.6

4)

x	-4	-3	-2	-1	0	1	2	3	4
y	-1	0	1	8	27	64	125	216	343

2a) 1c

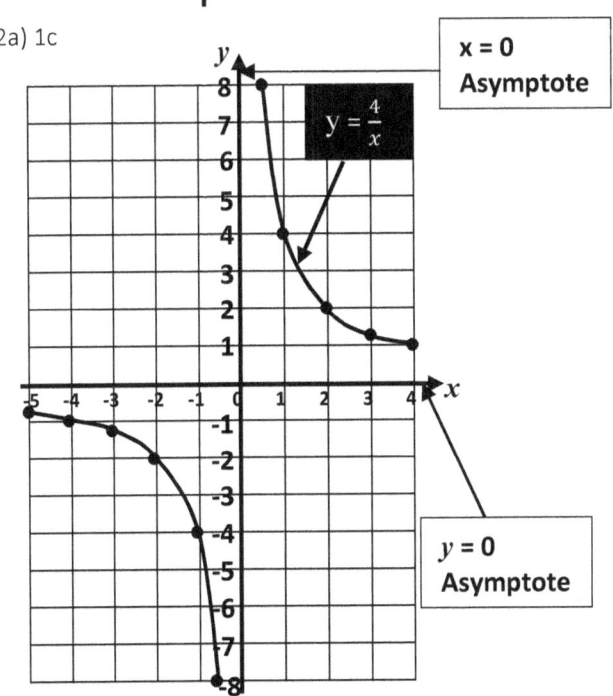

x = 0
Asymptote

$y = \dfrac{4}{x}$

y = 0
Asymptote

EXERCISE 11B

1a) $y = 3^x$

x	-4	-3	-2	-1	0	1	2	3
y	0.01	0.04	0.11	0.33	1	3	9	27

3a)

t	0	1	2	3	4	5	6	7	8
B	60	30	15	7.5	3.75	1.88	0.94	0.47	0.23

1b)

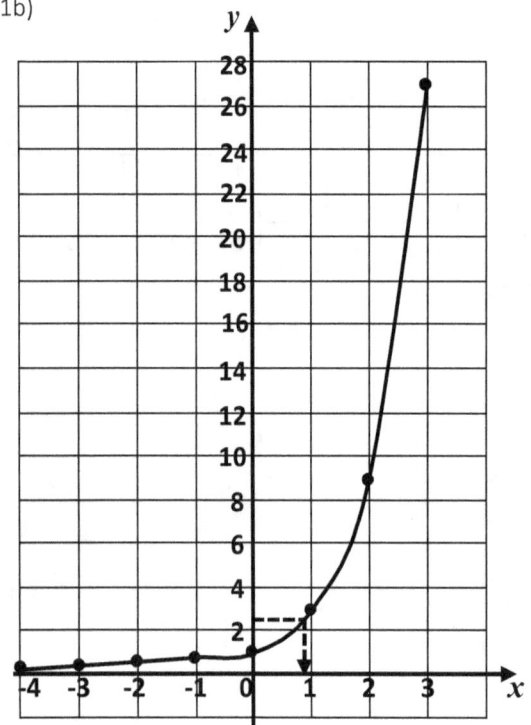

$x = 0.84$

3b)

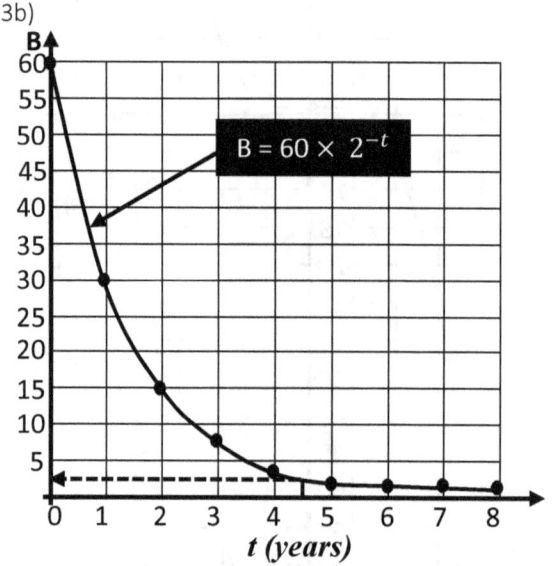

3c) Accept between 2 and 2.8.

4a) $y = (\frac{2}{5})^x$

x	-2	-1	0	1	2	3	4
y	6.25	2.5	1	0.4	0.16	0.06	0.03

4b) Student's own correct graph.

2a) $y = 5^x$

x	-3	-2	-1	0	1	2	3
y	0.008	0.04	0.2	1	5	25	125

2b) $y = 5^{-x}$

x	-3	-2	-1	0	1	2	3
y	125	25	5	1	0.2	0.04	0.008

2c) Student's correct graph.

EXERCISE 11C

1a)

θ	0°	30°	45°	60°	90°	135°	180°
$\sin\theta$	0	0.5	0.71	0.87	1	0.71	0
$\cos\theta$	1	0.87	0.71	0.5	0	-0.71	-1
$\tan\theta$	0	0.58	1	1.73	-	-1	0

1b)

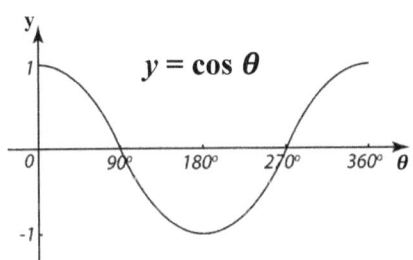

1c)

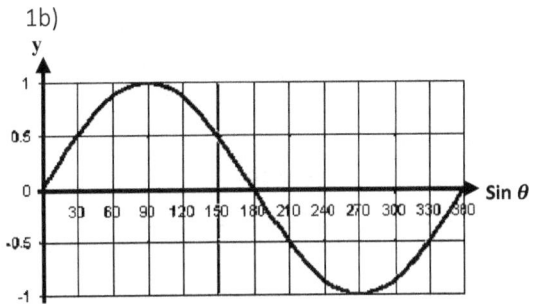

2) i) 30°   ii) 60°
3) i) 60°   ii) 30°

4a)

x^o	$2x^o$	$2\sin2x$
-180	-360	0
-135	-270	2
-90	-180	0
-45	-90	-2
0	0	0
45	90	2
90	180	0
135	270	-2
180	360	0

4b)

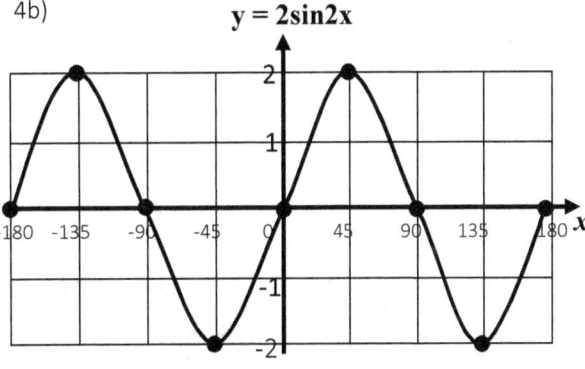

4c) Minimum values i) (-45, -2) and (135, -2)
Maximum values i) (-135, 2) and (45, 2)

EXERCISE 11D

1) x = 30°, 150°, -210° and -330°

2a)

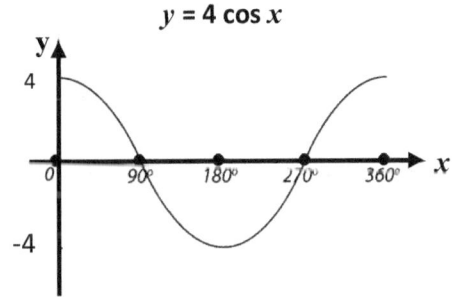

2b) x = 75.5° and 284.5°
3a) 37° and 143°
3b) 53° and 307°
3c) 348° and 192°
3d) 15° and 345°

4a) 11°   b) 11° and 169°

5a) i) -0.707   ii) 0.891
5b) 66.4°

EXERCISE 11E

1a)

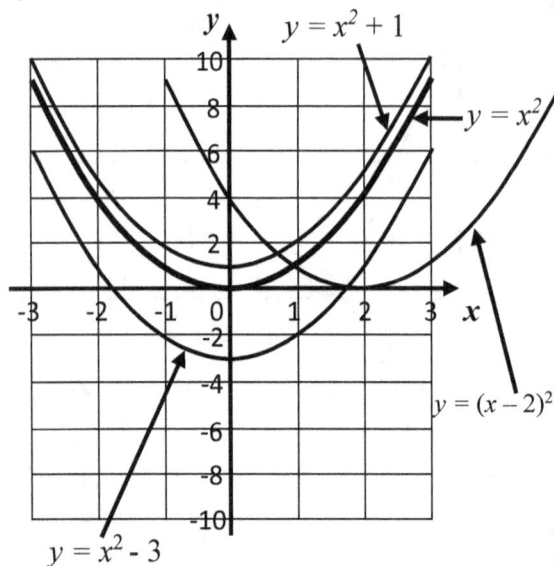

$y = x^2 + 1$
$y = x^2$
$y = (x - 2)^2$
$y = x^2 - 3$

1b) See diagram above

1c) i)

$y = x^2$ has been translated by the vector $\begin{pmatrix} 0 \\ 1 \end{pmatrix}$
to give $y = x^2 + 1$.

ii) $y = x^2$ has been translated by the vector $\begin{pmatrix} 0 \\ -3 \end{pmatrix}$
to give $y = x^2 - 3$.

iii) $y = x^2$ has been translated by the vector $\begin{pmatrix} 2 \\ 0 \end{pmatrix}$
to give $y = (x - 2)^2$.

2a)

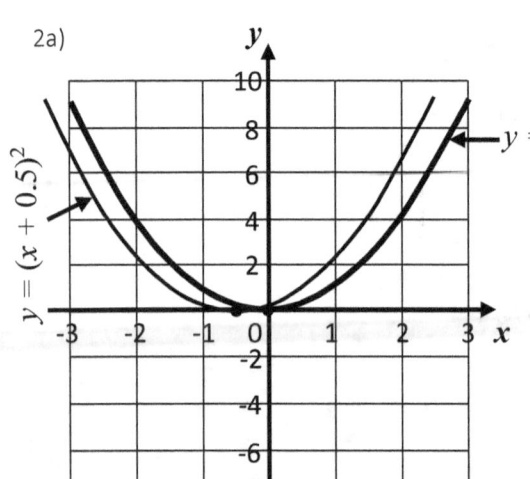

$y = (x + 0.5)^2$
$y = x^2$

2b) see diagram

2c) $y = x^2$ has been translated by the vector
$\begin{pmatrix} -0.5 \\ 0 \end{pmatrix}$ to give $y = (x + 0.5)^2$.

3a and 3b)

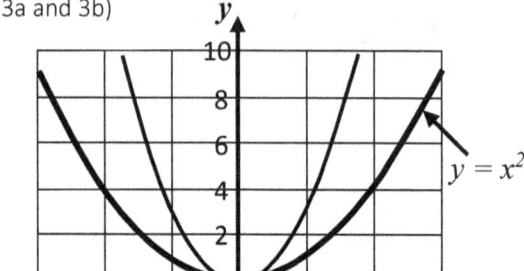

$y = x^2$

3c) The curve of $y = x^2$ has been stretched out in
the direction of the y-axis by a scale factor of 3
to give the graph of $y = 3x^2$.
Also, the y-coordinates of $y = 3x^2$ are 3 times
bigger when compared with the y- coordinates
of $y = x^2$.

4a) The curve of $y = x^2$ has been stretched out in
the direction of the y-axis by a scale factor of 6
to give the graph of $y = 6x^2$.
4b)
The x-coordinates of $y = x^2$ is a quarter
$(\frac{1}{2})$ compared to that of $y = (2x)^2$. It therefore
moves the graph of $y = x^2$ inwards (towards the
y-axis) to produce the graph of $y = (2x)^2$.

5)

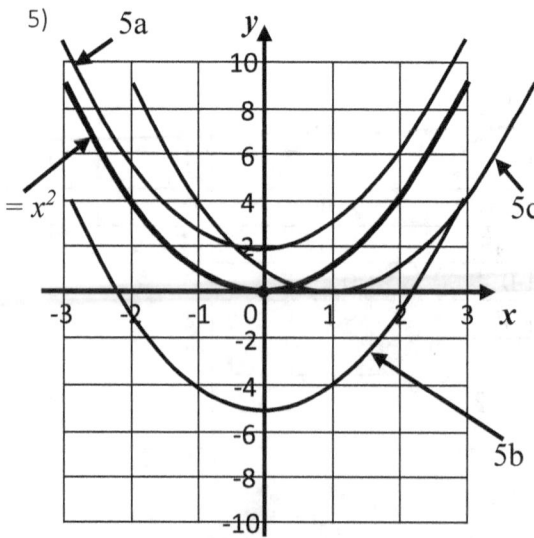

5a
5c
5b
$y = x^2$

6a) Translation 3 units down or by the vector $\begin{pmatrix} 0 \\ -3 \end{pmatrix}$.

6b) A stretch parallel to the y-axis by a scale factor of 3 and translation 2 units up by the vector $\begin{pmatrix} 0 \\ 2 \end{pmatrix}$.

Question 7a

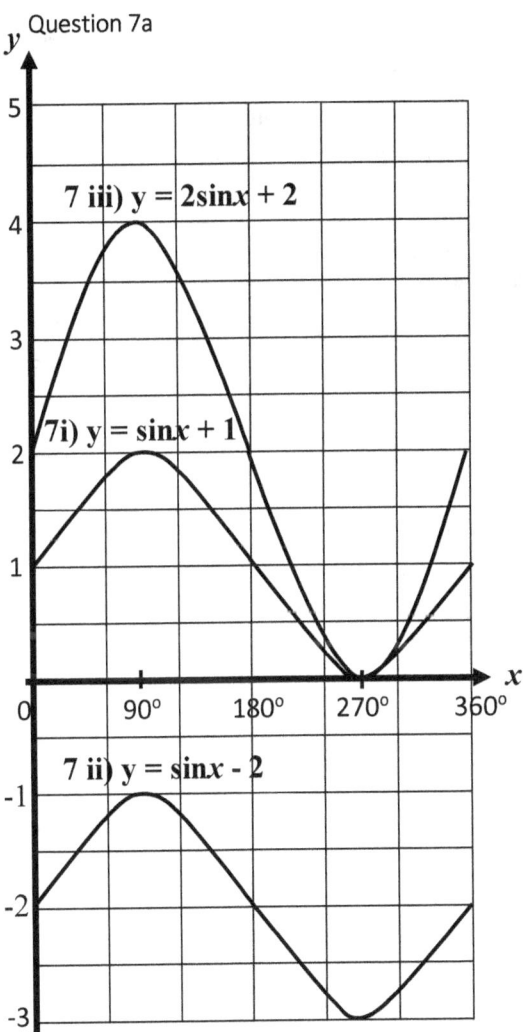

7 iii) $y = 2\sin x + 2$

7i) $y = \sin x + 1$

7 ii) $y = \sin x - 2$

Question 7b) The period is 360°.

107

INDEX